Björn Berg · Charles McClaugherty
Plant Litter

Springer

Berlin
Heidelberg
New York
Hong Kong
London
Milan
Paris
Tokyo

Björn Berg · Charles McClaugherty

Plant Litter

Decomposition, Humus Formation, Carbon Sequestration

With 76 Figures and 64 Tables

 Springer

Dr. Björn Berg
Department of Soil Ecology
BITOEK
University of Bayreuth
Dr. Hans-Frisch-Straße 1-3
95448 Bayreuth
Germany

Dr. Charles McClaugherty
Department of Biology
Mount Union College
1972 Clark Avenue
Alliance, OH 44601
USA

ISBN 3-540-44329-0 Springer-Verlag Berlin Heidelberg New York

Cataloging-in-Publication Data applied for

A catalog record for this book is available from the Library of Congress.

Bibliographic information published by Die Deutsche Bibliothek
Die Deutsche Bibliothek lists this publication in die Deutsche Nationalbibliografie;
detailed bibliographic data is available in the Internet at http://dnb.ddb.de

Springer-Verlag Berlin Heidelberg New York
a member of BertelsmannSpringer Science+Business Media GmbH
http://www.springer.de

© Springer-Verlag Berlin Heidelberg 2003
Printed in Germany

Cover design: Design & Production GmbH, Heidelberg
Typesetting: Camera-ready by the authors
31/3150WI- 5 4 3 2 1 0 - Printed on acid-free paper

Preface

When starting our work on this book we intended to summarize and synthesize the new information that had developed in the last 20-30 years in the field of plant litter decomposition. It turned out, however, that the main part of more recent work was directed towards boreal and temperate forest systems and therefore, with a focus on these ecosystems we finally concluded a synthesis that has a similarity to a case study. Still, we hope that a deeper insight into the behavior of a limited number of litter species will be of value for a generalization and also for the identification of process systems that deviate from those presented here.

We have written the book focusing on the transfer from newly shed litter to recalcitrant humus, describing and explaining the system of chemical changes taking place in the process both on a mechanistic basis and on a more general and regional level, considering different climates and species.

As a synthesis, this book gives some new aspects on decomposition that to some of us may be controversial. Thus, the fact that we emphasize the dominant role of microorganisms in the process may be disturbing to many readers, as well as the strong emphasis we give to the fact that humus layers actually do grow over millennia and that at a considerable rate, and thus really sequester e.g. C and N.

This book is based primarily on data and conclusions made from field studies. We have focused on undisturbed forest systems in an attempt to create a basic understanding and basic mechanisms for the decomposition and transformation processes. Its emphasis is on boreal systems for the obvious reason that there appeared to exist more data about these systems that could be synthesized. The information from temperate systems has rather supported and extended the conclusions, suggesting that the synthesis so far may be applicable to at least both types of systems.

In the topic of litter decomposition and transformations, we can not yet identify different schools of thought; it appears that this field of research has not yet developed far enough. We would rather consider different directions of the research work. Thus, some scientists have attempted to understand mechanisms for the degradation whereas several groups have searched for indices for prediction of long-term decomposition rates.

Many persons have helped us in the process of collecting and developing the information that makes up the backbone for this book. We want to thank all of them and hope they understand that all cannot be listed here.

The synthesis that we present has clearly taken impression of the work by a smaller group of scientists and research groups that we want to mention. Thus, the papers emerging from the group around Dr Marie-Madeleine Couteaux, CNRS,

Montpellier has been important to us, like those from Prof John Aber, University of New Hampshire, and Dr Jerry Melillo, The Ecosystems Center, MBL, Woods Hole. For the generalization to different climates the approach taken by Prof Vernon Meentemeyer, University of Georgia, Athens, has been very valuable.

We have also received extremely valuable help and direct support for the work. Thus, we want to thank Prof. Egbert Matzner, Director of the BITOEK Institute, University of Bayreuth for his extensive support for this book. The support of the Brumbaugh Center for Environmental Science, Mount Union College, Alliance, Ohio is also gratefully acknowledged. Financial support for part of this work came from the Commission of the European Union, through the CNTER project (QLK5-2001-00596) and finally the financial support from the company BBB-Konsult, Uppsala, is acknowledged.

Finally, before handing over the book to the reader we would like to thank each other for an excellent cooperation.

Bayreuth and Alliance, January 2003

Contents

1 Introduction

1.1 Overview of plant litter decomposition

Decomposition and photosynthesis are processes that account for a huge majority of the biological carbon processing on planet Earth. Photosynthesis has been studied extensively at levels ranging from biochemical to ecological. Relatively speaking, photosynthesis is well understood. Its importance to the functioning of the biosphere, as well as its agronomic significance, is well established. Furthermore, photosynthesis occurs in above ground tissues in organisms that are often large and aesthetically pleasing.

Decomposition accounts for the transformation of nearly as much carbon as does photosynthesis. However, it occurs mainly on or below ground. As such, decomposition is largely out of sight. It is carried out primarily by bacteria and fungi and is sometimes associated with products that are unappealing. The biochemistry of decomposition is very irregular when compared to the biochemistry of photosynthesis. Thus, it should be no surprise that decomposition is the less well studied of the two major carbon-transforming processes on the planet. In the last two decades, the need to have a better understanding of decomposition has become apparent. Decomposition of organic matter is responsible for huge amounts of the carbon dioxide returned to the atmosphere. It is also responsible for the formation of humic substances that contribute to soil fertility as well as the long-term storage of carbon. Decomposition is closely tied to nutrient cycling and is essential for the regeneration of organically bound nutrients.

Decomposition is more difficult to define than photosynthesis. Broadly defined, decomposition includes physical, chemical and biological mechanisms that transform organic matter into increasingly stable forms. This broad definition includes physical fragmentation by wet-dry, shrink-swell, hot-cold and other cycles. Animals, wind, and even other plants can also cause fragmentation. Leaching and transport in water is another important physical mechanism. Chemical transformations include oxidation and condensation. Biological mechanisms involve ingestion and digestion, along with extracellular enzymatic activity. Much of decomposition is ultimately accomplished by aerobic metabolism. The process is far from linear and many meta-stable products are created and accumulated during decomposition.

Scientists have long been familiar with negative exponential models. They describe the extinction of light intensity at increasing depths in a water column. They define the decline in radioactivity of a radioactive isotope over time. They have also been used to describe the loss of mass from decomposing organic matter in natural systems. As much as we would like Occam's razor to prevail, the simple linear decay function is not universally applicable to the decay of organic matter, though it often provides an excellent first approximation, especially during early stages of decay. Equation 1.1 is the basic exponential decay with time (t) as a variable

$$X_t = X_0 e^{(-kt)} \tag{1.1}$$

Unfortunately, neither the total biomass nor the constituent nutrients of decaying organic matter follow this simple function well enough to make it predictive. As a result, decomposition studies have taken two basic tracks. In one track, scientists have relied on mathematical abstractions and a variety of models, both theoretical and empirical to predict, or at least simulate, the decomposition process. In a second track, the intention has been to study decomposition experimentally in the field and in the laboratory, in an attempt to extract general principles of the process. Both approaches have merit, yet neither has been able to completely explain the complex process of decomposition.

Decomposition is so complex that it requires the attention of numerous scientific disciplines. Studies on litter decomposition and humus formation may encompass different branches of science and include, for example, chemistry, microbiology, climatology, geology, and systems ecology.

The objective of this book is to illustrate the importance of decomposition as a major ecological process and to review and summarize the current understanding of the decomposition processes in natural systems. We rely heavily on work that has been done in temperate and boreal forests, but we also refer to other ecosystems.

1.2 A short retrospective

In 1929, Tenney and Waksman postulated that decomposition rates of soil organic matter are controlled by four distinct factors; (1) the chemical composition of the substrate; (2) a sufficient supply of nitrogen for the decomposer organisms; (3) the nature of the microorganisms involved; (4) environmental conditions, especially aeration, moisture supply, and temperature. As a general statement concerning factors that control decay, this is still valid more than 70 years later.

During the intervening 70 years, we have begun to understand many of the details of decomposition. We have gained a deeper insight into the effects of substrate quality and environmental factors, both on a local and regional scale. Studies have begun to show us more about the implications of decomposition on nutrient cycling. We have learned something about the microbial enzymes involved in degrading complex organic substrates. Moreover, we have begun to think of decomposition as more than just the breakdown of organic matter, but also as the formation of stable humic substances.

In the 1970s, several scientists reviewed the state of knowledge relating to plant litter decomposition. Singh and Gupta (1977) reviewed studies on plant litter decomposition with a focus on soil respiration. Schlesinger (1977) approached decomposition from the perspective of carbon balance and brought together a large body of literature that describes carbon pools and fluxes in ecosystems of the world. His review emphasized the importance of long-term carbon storage in soils. At about the same time, Swift, Heal and Anderson (1979) produced a book that reviewed what was then known about decomposition in terrestrial ecosystems.

Mindermann (1968) was among the first to challenge the idea of simple negative exponential kinetics. He proposed that as litter decomposes, the readily decomposable materials would disappear first, leaving the more recalcitrant substances behind. As a result, decay should occur at a decreasing, rather than a constant rate. As we will see, even this approach is not entirely correct, but it is closer to reality than the simple, constant rate, loss model.

Nitrogen and lignin were soon recognized as major variables that influence the rate and pattern of decomposition (Fogel and Cromack 1977; Melillo et al. 1982). During the 1980's, numerous investigators focused on the role of these two compounds in regulating the decomposition process. These studies laid the foundation for our current understanding of the decay process and are described extensively in subsequent chapters.

1.3 The ecological significance of litter decomposition and the formation of humus

At the biosphere level, an understanding of decomposition is important for two main reasons. First, significant amounts of carbon dioxide, methane, and nitrogen-based gases are released as products of decomposition. These so-called greenhouse gases are of great interest currently because of their roles in potential global climate change. Factors that increase the rate of decomposition could serve to increase the amounts of carbon-based gases in the atmosphere. Second, soils represent a major sink for carbon. To the extent that carbon is stored in soil as humus and related stable organic compounds, it is not being circulated through the atmosphere. Thus, an understanding of the factors influencing the amount of humus formed and the stability of that humus are also important in predicting global atmospheric carbon budgets (Schlesinger and Andrews 2000).

At the ecosystem level, decomposition is important for somewhat different reasons. Nutrient cycling is clearly related to decomposition. The availability of nutrients in a given soil is due in large part to the decay dynamics of the organic matter in that soil. In addition, the accumulation of organic matter in soil can greatly increase the cation exchange capacity and have positive impacts on the nutrient holding capacity of that soil. Decomposition can influence the pH of soil; pH may be increased if plants pump basic cations up from the mineral soil to be released during the leaching and decay of litter. Soil pH can be lowered through the release of CO_2 and the formation of carbonic acid. Finally, during initial stages of decay, nutrients are immobilized, taken out of the general circulation for a while, thereby temporarily reducing nutrient availability.

Another impact of decomposition is on diversity and stability of the ecological community. Entire food webs are based on decomposition. In fact, detrital food webs process more carbon and energy than do the better-known grazer food webs. Only a small amount of primary production is grazed by herbivores and passed to higher trophic levels. In contrast, all ecosystem production ultimately becomes detritus.

The ecological significance of decomposition and humus formation can also be viewed from the perspective of microbial ecology. The decomposition of the litter substrate can be divided into phases, in which different parts of the substrate give energy to different groups of the microflora. This process supports diversity in the microbial population by supplying a rich set of intermediate degradation products and serving as energy and nutrient sources for different microbial subpopulations.

Decomposition influences other soil processes. For example, part of the litter forms humus and organic acids which in part are responsible for weathering in the mineral soil, thus supporting the supply of plant nutrients. Humus may also act as a carbon source for microorganisms, which subsequently produce acids that contribute to weathering. Nutrients are stored in humus. In practice, this creates a reserve of nutrients for plants that may be mobilized through a variety of mechanisms, such as fire or by stress signals from the trees to their mycorrhizae.

In addition to weathering, decomposition and humus formation are involved in storage and controlled release of nutrients to the plant and microbial communities, as well as the storage dynamics of carbon compounds. Further, the decomposition of litter and humus may also produce precursors for pathways other than the common one that leads to carbon dioxide. Well-known examples would be the fermentations that produce methane and the ones that produce organic acids.

1.4 Factors influencing decay and humus formation

Decomposition begins with complex plant detritus and produces carbon gases and humus. The process can be characterized by the rate of mass loss and the rates of nutrient immobilization and release. In addition, the chemical composition of decaying litter changes during decay. These changes are not, in all cases, linearly associated with mass loss. Neither are the changes in composition the same for similar litter substrates decomposing under different environmental conditions. Thus, there is a complex and interacting set of factors that regulate mass loss, humus formation, nutrient dynamics and patterns of change in chemical composition of decomposing plant litter.

Factors that influence either rates or patterns of decay will be introduced; litter chemical composition, climate, nutrient availability, communities of soil organisms and site-specific factors. The relationship between human activities and decomposition will be presented. Although there is an emphasis on boreal and temperate ecosystems reflecting the relatively large amount of data from these systems, we have attempted to create a basic image for decomposition and dynamics of plant litter/ soil organic matter (SOM).

The chemical composition of litter differs among plant species and tissue type, thus creating unique environments both as regards the larger chemical components and nutrient composition. The importance of these substances in regulating decay varies among litter types, under different climates and on different sites. These possibilities increase drastically when we consider a wider range of nutrients (e.g. N, P, S, K, Ca, Mg, Mn, Cd, Pb, Zn, Cu). In addition, litter pH may have a regulating effect, directly on the microorganisms involved in the decomposition processes and indirectly on the solubility and thus availability of the individual nutrients. Nikolov and Helmisaarii (1992) listed 58 dominant tree species for boreal Eurasia and North America. Of these, 14 species dominate in Fennoscandia and Russia/Siberia and 15 in North America. We know the chemical and nutrient composition for only a few of these species, so that it is difficult to generalize to entire forest communities.

1.5 Accumulation of humus and nutrients

The nature, rate and amount of carbon storage vary greatly among ecosystems. We know, for example, that anaerobic environments conserve their carbon compounds in peat or humus and we normally find more carbon stored there than in drier environments. Thus, anaerobicity may be one factor, but is it the dominant one? Could the chemical composition of the litter also have an influence? Even among aerobic systems with similar vegetation, we can find varying amounts of stored humus from almost bare mineral soil with about 200 to 300 g of humus m^{-2} on top of the mineral soil to layers nearly 150 cm thick and with ca. 50 kg of humus m^{-2}. Therefore, there may be mechanisms for long-term humus accumulation that can explain the wide range of accumulated humus. Disturbances such as fire, harvesting for forestry or agriculture and cultivation can clearly reduce soil organic matter content. A natural question is thus; what could such a mechanism be, is there just one or does a series of coincidences rule whether and to what extent humus will accumulate?

A humus layer of ca. 50 kg of humus m^{-2} accumulated over ca. 3000 years (Wardle et al. 1997) means, on average, an annual accumulation of ca. 17 g of humus m^{-2}. Such a powerful mechanism for storage of C could mean a long-term storage of 0.1 to 0.3 Pg (Petagram = 10^{15} g) $year^{-1}$ considering the whole boreal forest zone of 1-2 billion ha (UNECE/FAO 2000), which accounts for 10 to 20 % of the currently estimated 1-2 Pg unidentified sink for carbon (Woodwell et al. 1998; Houghton 2001).

There is often a rather resistant part of the litter remaining, a part that may be stored on a long-term basis (e.g. millennia) and be regarded as humus. This long-term storage encompasses both carbon and nutrients. For carbon, amounts of the magnitude of 50-108 kg of humus m^{-2} have been found in the form of organic layers (Berg et al. 1993c; Wardle et al. 1997). For nutrients such as N and P (see Chap. 11), amounts of 760 and 39 g m^{-2}, respectively, have been measured. A long-term storage and in such amounts raises doubts about the concept of a steady state as suggested by a number of authors, such as Schulze et al. (1989).

Plant litter contains nutrients at differing concentrations and is thus a carrier of nutrients that are largely released when the organic matter is being decomposed. The release of nutrients may take place in different ways depending on the type of litter. Foliar litters may leach nutrients with especially mobile ones such as K being leached to a high extent. Other nutrients may be partially leached, a release that may depend on their concentrations in litter. The major nutrients, such as N, P, and S, can be partially leached immediately at litter fall. This immediate leaching may be followed by an accumulation or a net uptake of nutrients to the litter with a later net release. The release of such structural nutrients is often in proportion to litter mass loss and thus regulated by the same factors that regulate the rate of litter mass loss. Normally all nutrients are not released. Available studies indicate that some components, such as most heavy metals, are not released from litter, even in clean, unpolluted systems, but rather accumulate so that when the decomposition process becomes extremely slow, close to the humus stage, the remaining part has a rather high concentration of such nutrients.

The boreal forest has a multitude of different storage forms of carbon, in the form of dead organic matter. Not only as numerous different chemical compounds of which we will never know more than a fraction, but also on a large scale, in the form of different ecosystem niches. On an even larger scale, we may distinguish between living forests with soil systems that are more or less aerobic and more or less anaerobic. Both soil system types would be expected to store carbon on a long-term basis unless disturbed by, for example, fire, ditching, or site preparation.

1.6 The contents and organization of the book

This book brings together much of the current understanding of the decomposition process as it occurs in forested ecosystems, using examples primarily from boreal and temperate forests. The book begins with a presentation of decomposition as a process (Chap. 2). Terminology related to decomposition is briefly introduced including litter, humus, mineralization and immobilization (also see Glossary). The overall process of decomposition is reviewed from the input of fresh litter through the dynamics of microbial and physical decay to the formation of meta-stable humic substances. An examination of litter quality includes a presentation of the cellulose, hemicellulose and lignin structures and the enzymes attacking them. A simple conceptual model is introduced that illustrates the decomposition processes and identifiable functional steps. The model, which is used as an organizing principle, encompasses three stages of litter decomposition with contrasting functional properties.

The book then moves to examine the biological agents of decay (Chap. 3). The most important groups of decomposer organisms are fungi and bacteria, which in boreal coniferous forests may be responsible for more than 95% of the decomposition. Their importance in deciduous forests is somewhat less (Persson et al. 1980). Even in tropical forests and grassland communities where herbivores may consume larger proportions of net primary production, they are notably inefficient and microorganisms are still the major decomposers. The chapter, therefore, has an emphasis on the microflora. This chapter emphasizes functional roles of organisms rather than their taxonomy. The main groups of fungi and some bacterial genera will be mentioned but with an emphasis on function and activity.

With this biological background, the next two chapters (Chaps. 4-5) examine the importance of initial and changing chemical composition of decaying litters. The initial substrate for decomposition is newly shed litter. Its chemical composition determines both the composition of the microbial community and the course and pattern of the decomposition process. Both the organic and the nutrient composition of litter vary with species and with climate and site properties. Generally, deciduous foliar litters are richer in nutrients than coniferous ones. Organic components are also highly variable; lignin concentration may vary from a few percent in some litters, up to about 50% in some others.

As the decomposition process proceeds, chemical changes take place in the decomposing litter such as increasing concentrations of some nutrients and most heavy metals. The concentration of the organic component lignin increases too, whereas the changes in concentrations of the main carbohydrates vary. A basic pattern of chemical changes and an overview of how such changes vary depending on the initial chemical composition, are discussed.

In Chapter 6, the attention is shifted to the influence of changing litter quality on the decay processes. This chapter shows that the influences of selected litter components change dramatically during the process, sometimes even reversing the direction of the effect. For newly shed litter, differences in initial N, P and S levels influence the rate of the decomposition process. In a later stage of decay, the increasing lignin levels suppress the decomposition rates in ways that interact with N concentration and climate. Finally, the decomposition process normally reaches a stage at which decomposition goes so slowly that the stage may be approximately described by an asymptote or a limit value. Factors that influence this limit value are reviewed and evaluated.

Decomposition occurs in a natural environment where climate (Chap. 7) and other site factors (Chap. 8) can have a profound influence on the decay dynamics. Climate is a powerful regulating factor for litter decomposition as shown on a larger scale using actual evapotranspiration (AET; Meentemeyer 1978) and will directly influence litter decomposition rate, especially that of the newly shed litter. This effect of climate will decrease as decomposition proceeds. We will use the conceptual model introduced in Chapter 2 and developed in Chapter 6 to describe how the effects change with the phase of decomposition. The other major environmental factors that influence decomposition are edaphic; soil texture, nutrient availability and soil chemistry. For example, a site factor that may have an influence is the parent rock material. Granite parent material, being nutrient poor, gives a poor litter substrate, whereas one of limestone often increases the nutrient content of the litter. The texture of the soil (sand and clay content), influences nutrient mobility and hydrology, which can affect the decomposition rate both indirectly and directly.

Some litter types, notably wood and roots, behave differently from foliage and these substrates require treatment in a separate chapter (Chap. 9). Woody debris and roots may represent a large proportion of the total litter input into an ecosystem. The decay of these substrates is often very different from that of foliar litter and the course of the decomposition process is directly dependent on whether the wood is attacked by white-rot or brown-rot fungi. Although much less data is available on these litter types than for foliar litter, it appears that the attack by white-rot or brown-rot fungi is an important factor for determining both the rate and course of decomposition and for the long-term accumulation of soil organic matter.

As knowledge of decomposition has grown, scientists have created a number of models to describe or predict decay (Chap. 10). These models vary in intent and complexity. Simple mathematical models have been used, (e.g. a linear one and simple exponential ones) that can describe the process, including rates of decomposition, and which are in agreement with conceptual models. Four such models are discussed; single exponential, double exponential, an advanced triple exponential and an asymptotic model and are evaluated with regard to recent findings.

Many ecosystem properties and processes are closely related to decomposition and these are briefly reviewed in Chapter 11. Topics closely related to decomposition include nutrient cycling, soil organic matter formation and carbon sequestration and efflux. A major rationale for studying decomposition is the strong link to carbon and nutrient cycling and storage. Large amounts of nutrients may be stored in the soil organic matter of an ecosystem, both as integral components of the organic residues and as attached ions on the organic colloids. In addition, the role of humus in soil structure and function as well as the potential importance of forest soils for long-term carbon storage are discussed.

The book concludes with a survey of how human activities have influenced decomposition processes (Chap. 12). Ecosystems will respond differently to climate change. For example, a change in humus accumulation is predicted for part of the taiga belt. Forestry and agricultural practices, including fertilization and site preparation, will cause changes in the decomposition process.

The terminology used in decomposition studies is sometimes confusing or even misleading. In order to make our text as clear as possible we have provided a glossary that explains and defines terms used in this book. We have also provided an appendix that gives Latin names together with English names for all vascular plant species that we discuss. We have used the plant names as they were given in the original articles. As some species have different common names in American and European English, we have given both names.

1.7 Motives for the present synthesis

We hope this book will help aim research efforts towards furthering our understanding of decomposition. During the past several decades, numerous investigators have studied many aspects of the decomposition process. At the same time, both scientists and the public have become concerned about the increasing concentrations of carbon gases in the atmosphere and the implications for global climate change. As a result, we felt the need to compile existing recent information on boreal forest litter and humus with the aim of giving a new basis for understanding what regulates the buildup and stability of humus. Much of the book focuses on the processes and controls of the early stages of decay, but these early stages influence the long-term carbon dynamics of each ecosystem. A number of basic questions are raised, including:
- Is there a long-term buildup of humus?
- Do forest systems exist with very high or very low buildup rates?
- Can we influence the humus buildup rate?
- Are there any large-scale threats to the natural course of the process?

This book is based primarily on data and conclusions made from field studies. We have focused on undisturbed forest systems in an attempt to create a basic understanding and basic mechanisms for the decomposition and transformation processes. The book focuses on boreal systems for the obvious reason that there were more data about these systems available for analysis. Of course, we have used information from temperate systems when applicable and there have not been any conflicting conclusions, suggesting that the synthesis so far may be applicable to both kinds of systems.

When it comes to the topic of litter decomposition, we have not identified different schools of thought; as it appears that this field has not yet developed far enough. We would rather consider different directions of the research work. Thus, some scientists have attempted to understand mechanisms for the degradation whereas several groups have searched for indices for prediction of long-term decomposition rates.

2 Decomposition as a process

2.1 Litter decomposition - a set of different processes

Decomposition of plant litter involves a complex set of processes including chemical, physical and biological agents acting upon a wide variety of organic substrates that are themselves constantly changing. Due to the immense diversity of possible factors and interactions, decomposition in a natural setting can be described in general terms only. In spite of this complexity, several major processes are involved and general trends can be outlined.

Litter is in its simplest state when shed by the plant. From its initial state, litter composition changes, with some litter components disappearing rapidly, some slowly and some beginning to disappear only after a time delay. Perhaps non-intuitive, but very significant, is the fact that some substances, particularly nutrients, are imported into the decomposing substrate and new organic compounds are synthesized during decomposition. Due to the heterogeneity in both litter composition and factors influencing decay, decomposition of litter is far more complex than decay of, for example, a radioactive isotope.

The bulk of plant litter consists of varying amounts of several major classes of organic compounds. The relative proportions of these compounds vary with plant part (for example, leaves, stems, roots, bark) and among species (see Chap. 4). These major groups of compounds can be classified according to their molecular size, their solubility and their primary constituents. Some materials, notably sugars, low molecular weight phenolics and some nutrients, are readily lost from litter through dissolution and leaching combined with the action of rapidly growing opportunistic microorganisms. Larger macromolecules, including cellulose, hemicelluloses and lignin are degraded more slowly. During decay, condensation of phenolics and lignin degradation products, combined with the import of nutrients, results in the net accumulation of newly formed substances. The relative magnitudes of the main flows (Fig. 2.1) are thus different not only among litter types but are influenced by litter chemical composition (see Chap. 6).

We regard "litter mass loss" or "decomposition" as the sum of CO_2 release and leaching of compounds, including both C compounds and nutrients. Leaching is simply the loss of nutrients and incompletely decomposed organic compounds transported out by water from the remains of decomposing litter (see Glossary).

The interpretation of mass-loss data during the initial stages of decay may be influenced by a high leaching rate of water-soluble material that is not physiologically modified by microorganisms (McClaugherty 1983). These dissolved materi-

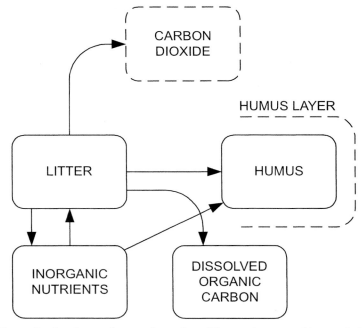

Fig. 2.1. Generalized pathways for transformation of litter to humus and inorganic C. When the litter is shed and its decomposition has started, the microorganisms begin forming carbon dioxide, and soluble compounds that are initially present may be leached out. Newly formed compounds that are stable but water-soluble are also leached out (dissolved organic carbon - *DOC*), and long-term stable remains form humus

als may be lost from litter and subsequently sequestered by humus or clay particles. In such cases, the materials are lost from a particular substrate but are retained in the soil ecosystem.

Under aerobic conditions, microbial decomposition results in a release of CO_2 that leaves the system. Under more anaerobic conditions, such as a temporarily waterlogged organic-matter layer, anaerobic decomposers may produce organic acids instead of CO_2. This may also happen with aerobic decomposers that suffer from a lack of oxygen. For example, acetic acid may be released instead of CO_2 and either be decomposed outside the cell or be stored and fulfill another role (see Chap. 11).

In later stages of decomposition, recalcitrant substances become more abundant in the residue and in some cases the rate of decomposition approaches zero. In 1974, Howard and Howard estimated limit values for the decomposition of some kinds of leaf litters that were incubated in an animal free environment. Using litter decomposition data from nutrient poor forest systems, Berg and Ekbohm (1991) also estimated such limit values, indicating a stage at which the decomposition rate nears zero (Fig. 2.2).

This chapter describes and explains the principal microbial processes associated with litter decomposition that result in mass loss or CO_2 release. We present basic principles about the microbial degradation of the main components of litter such as cellulose, different hemicelluloses and lignin. The process of litter mass loss is described in some detail for one litter type, namely Scots pine needles. We organize our description using a conceptual model that separates the decomposition process into different phases. As litter passes through these phases, the factors that regulate the decay process change. The model connects the developmental processes that occur, beginning with newly formed litter and continuing to the formation of humus. Differences in the importance of substrate properties (for example, the influence of N and lignin) as the decay process unfolds are emphasized.

2.2 Definition of litter decomposition

Litter decomposition may in part be defined by the method used to study it. A very common method is the litterbag, used for incubations in the field or in laboratory microcosms. Another variety of direct incubation is tethered litter. With these kinds of measurements, decomposition is measured as loss of mass and studies normally do not distinguish between what is respired as carbon dioxide and what is leached out of the litter or lost due to fragmentation, unless those processes are investigated separately. In terms of mass, mobile nutrients such as K and Mn are also part of the loss. Ingrowth of microbial biomass and the transport of nutrients into the litter, result in the movement of mass into the litter that was not there originally. Thus what is often called "litter mass loss" or "decomposition" is a *net mass loss* although the ingrowth of mycelium normally is negligible from the point of view of mass. As we will see later, the *in situ* incubation of intact litter is, from several points of view, preferred to laboratory incubation methods but it still may be compared with laboratory studies.

When litter decomposition is measured as respiration, only part of the mass-loss process is quantified. No specific term has been suggested for this more specific process, but terms such as "release of CO_2", "C- mineralization" and "litter respiration" are used. We will use "litter CO_2 release" (see Glossary in Appendix I) in this book. Thus, the two processes of "litter CO_2 release" and "leaching" together, should correspond to "decomposition" as the term is used today (Fig. 2.1). Distinguishing litter-based respiration from other respiration in the field is difficult. Methods are needed to separate other sources of respiration, for example, root and faunal respiration that is not directly associated with litter decomposition.

In boreal forest systems, microorganisms carry out more than 95% of the litter decomposition (Persson et al. 1980). Before litter falls, some microorganisms are present on the litter. Most of these are not involved in decomposition unless they are pathogens. After litter fall, fungi are the first invaders, penetrating the leaf through openings and thus invading the fresh substrate. The less mobile bacteria come later and there is also a succession of fungal species with different physiological properties depending on the decomposition stage and thus the substrate quality of the litter.

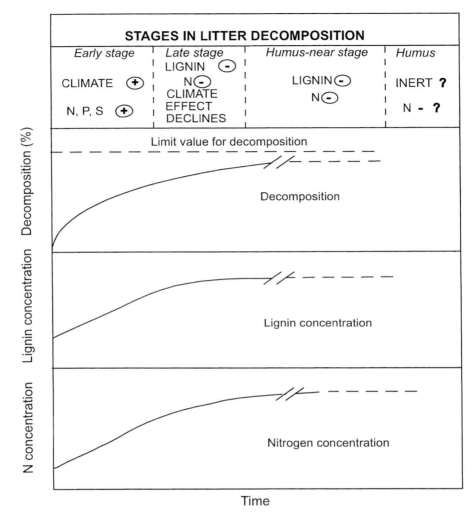

Fig. 2.2. Stages in litter decomposition. We may distinguish three phases before humus is formed (*top*) that have different properties from the point of view of decomposition (and CO_2-release). In the early stage, decomposition of solubles, and unshielded cellulose and hemicelluloses takes place, in a process influenced by climate. In the late stage, the influence of climate on decomposition gradually decreases to nothing. In the same late stage, N may have a negative influence on lignin degradation through a repression of *de novo* ligninase synthesis and/or through creating a barrier based on chemical bonds between lignin remains and N. Finally, in a humus-near stage, the decomposition reaches a limit value. From the onset of decomposition, the concentrations of N and lignin increase. The + and – signs indicate positively and negatively related effects, respectively, to increased concentrations of nutrients and lignin

2.3 Degradation of the main groups of compounds in litter

2.3.1 Degradation and leaching of soluble organic substances

Foliar litter may contain considerable levels of soluble substances. For example, concentrations of water-soluble substances between 7% in Scots pine needles and 30% in grey alder leaves have been recorded (see Chap. 4). Part of these substances may be leached out of the litter (Bogatyrev et al. 1983; McClaugherty 1983) and part may be degraded in the litter structure. There are four principal groups of soluble organic material in litter: sugars, phenolics, hydrocarbons and glycerides. The soluble sugars are predominantly mono- and oligosaccharides that were involved in metabolic processes of the plant. The soluble phenolics are low molecular weight compounds that serve either as protectants against herbivory, are lignin precursors, or are waste products; tannins are a common example of soluble phenolics. Phenolics are highly variable in their solubility and many have a tendency to condense into less soluble forms or to react with larger molecules. Not all nutrients are soluble; many are bound into organic complexes.

2.3.2 Patterns of degradation of the main organic compounds in litter

Decomposition generally follows a sequential pattern with different classes of organic compounds dominating the decay process as it proceeds. In general, degradation of soluble and low molecular weight compounds dominate the first stage of litter decay. Next, hemicelluloses, especially those that are arabinan-based, begin to disappear. Somewhat later, cellulose degradation is the dominant activity and finally lignin degradation becomes dominant. Although individual processes may dominate a particular stage of decomposition, any or all the processes may occur to some extent throughout the decay continuum (Figs. 2.3, 2.4; Table 2.1). This general pattern has been observed in numerous studies. Aber et al. (1984) described the patterns for 12 types of forest litter. One early interpretation was that there are different decomposition rates for different chemical components (Mindermann 1968), an opinion that is not quite correct.

In a study on Scots pine needle litter the degradation of single components was followed (Berg et al. 1982a, Fig. 2.3). The onset of mass loss due to polymer carbohydrate degradation, namely cellulose and single hemicelluloses, began after different periods of incubation. The degradation of arabinan and galactan begins almost immediately after litter fall (Table 2.1). Degradation of cellulose and mannan starts later, with the first measurable loss of these components being observed after ca. 1 year. Figure 2.4 shows that the amount of cellulose increases before a decrease starts and a net decrease does not start until after about one year. For xylan, the degradation started about half a year later, while with lignin, the first net loss was noted after 2 years (Figs. 2.3, 2.5).

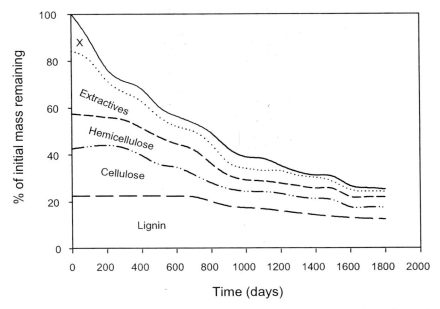

Fig. 2.3. The disappearance of the main organic-chemical components from Scots pine needle litter decomposing in a boreal Scots pine forest. *X* denotes unidentified compounds. From Berg et al. (1982a)

The degradation pattern is related to the arrangement of the components in the fiber (Fig. 4.1). Microorganisms may first attack and degrade those carbohydrates that are located on the more available outer structures. A practical consequence of this is that after the onset of decomposition, the relative degradation rates of the different components are initially different (Table 2.1; Fig. 2.5). Figure 2.5 illustrates the typical pattern. When degradation of lignin begins there may be little or no easily available carbohydrate and the microbial community must change to one that degrades lignin. In this environment, namely in needle litter with an N concentration above the initial level of 4 mg g^{-1}, there may be a suppression of lignin degradation (cf. Sect. 6.2.3) and in this study its rate was less than half of the initial rate of carbohydrate degradation. The rate of lignin degradation from the start of year 2 until the end of year 5 was about 0.04% day^{-1}. When we compare the degradation rates of the carbohydrates during the same period they were similar in magnitude to that of lignin.

This suggests that these components are degraded at the same rates because they are so well mixed in the fiber structure that they cannot be degraded separately. It thus appears that we may see at least two different groups of carbohydrates: those for which degradation starts immediately after litter fall (hemicelluloses dominated by arabinans and galactans); the second made up of mannans, cellulose and xylan for which degradation starts later. This could mean that the second group of components is less available than those in the first group as a

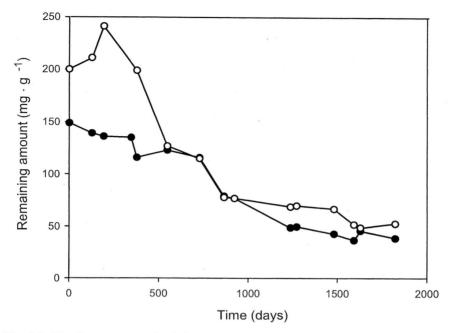

Fig. 2.4. The disappearance of cellulose (○) and hemicelluloses (●) from Scots pine needle litter decomposing in a boreal Scots pine forest. (Berg et al. 1982a)

Table 2.1. The time for onset of net mass loss of different organic-chemical components and their relative degradation rates in decomposing Scots pine needle litter. The rates given refer to two stages of decomposition, namely the early stage when the unshielded polymer carbohydrate components and solubles are degraded independently of lignin and the late stage when the degradation of carbohydrate components are regulated by the degradation of lignin and all components have similar rates (cf. Fig 2.5). Data from Berg et al. (1982a)

	Lignin	Cellulose	Mannans	Xylans	Galactans	Arabinans
Onset [days]	726	376	376	545	Immediate	Immediate
Rate [% day^{-1}] (0-726 days)	–	0.1041	0.0647	0.1077	0.0633	0.1461
Rate [% day^{-1}] (> 726 days)	0.0418	0.0393	0.0526	0.0461	0.0375	0.0449

result of being more lignified. In addition, a hemicellulose like arabinan is easily hydrolyzed at a lower pH (O. Theander, pers. comm.).

Few attempts have been made to follow the degradation of simple soluble components in litter and it should be pointed out that most studies describe net disappearance only. The soluble fraction is very challenging to study, due to the complexities of tracing the formation of new solubles during decomposition and the disappearance of the same solubles due to leaching or metabolism. For example, glucose, which is present initially in newly shed litter is also produced from decomposing cellulose and thus is found even in the later stages of decomposition.

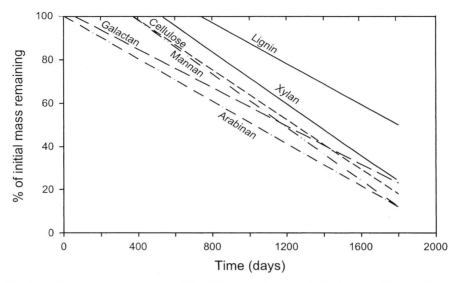

Fig. 2.5. Onset of decomposition of the different polymer carbohydrates and lignin, showing the different rates of decomposition in the initial and the late phase. Decomposition rates are given in Table 2.1. Data from Berg et al. (1982a)

The same applies to the simple sugars of hemicelluloses. Also several phenolic substances are found in newly shed litter, but are also produced during the degradation of lignin.

Berg et al. (1982a) found that compounds such as triglycerides and hydrocarbons that are soluble in light petroleum disappeared quickly whereas fatty acids and diterpene acids remained. Compounds such as simple sugars, for example, glucose and fructose, or ones related to simple sugars such as glycosides and pinitol, were also degraded very early and at a high rate (Fig. 2.6).

2.4 A model for decomposition from newly-shed litter to humus-near stages

In the section above we demonstrated that the decomposition patterns for organic-chemical components were different in the early stage as compared to later ones. We can use information obtained from pure-culture and physiological studies on microorganisms, compare this to the degradation of different components in our case-study litter and organize a conceptual model (Fig. 2.2). In the model, explanations are given for the different steps, which are connected to *in situ* decomposition experiments for litter and humus. Although each stage can be uniquely described, the process is more accurately a continuum in which transition points cannot be defined precisely.

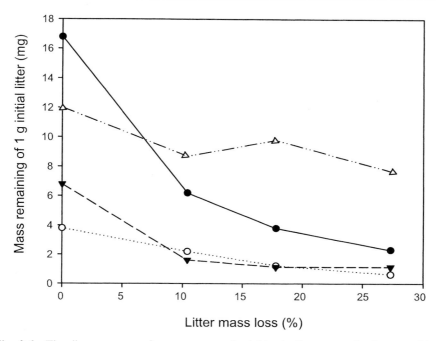

Fig. 2.6. The disappearance of some groups of solubles in the very early decomposition phase. Combined simple sugars (●), hydrocarbons (○), triglycerides (▼), acidic compounds (Δ). (Berg et al. 1982a)

The model describing the decomposition of litter towards humus may be divided into functionally defined stages: (1) newly shed litter - early decomposition stages; (2) late stages for partly decomposed litter; (3) humus-near stages or limit value stages where litter is close to becoming stable humus. These are the main stages that we intend to connect. The model is focused towards the effects of climate, the effects and roles of nutrients in the early phase and of lignin and N in the late phase, but also accounts for theories for humus formation (humus-near stage, above). As an introduction we briefly explain the essential contrasts between the effects of N in newly shed litter and in litter that approaches humus.

The reasons to divide the decomposition process into different stages are rather straightforward. The substrates, for example needles or leaves, that microorganisms decompose are far from homogeneous. On the level of plant cells the polymers lignin, cellulose, and hemicelluloses are structurally organized (see Fig. 4.1) and the main part of the cellulose and hemicelluloses are found in the primary cell wall, whereas lignin is distributed in the secondary wall and in the middle lamella (Eriksson et al. 1990). A result of the distribution in the cell wall is that there is a separation of carbohydrates into those that are not lignified and those that are encrusted into lignin. Microorganisms that are not lignolytic may degrade only the former.

2.4.1 The early decomposition stage: degradation of solubles and non-lignified carbohydrates

The ingrowth of microorganisms, mainly fungi, into the litter, may begin prior to litter fall, but the ingrowth of decomposers takes place when the litter has reached the ground. The more fast-growing microorganisms start invading the litter, with part of the litter C becoming microbial biomass and part CO_2. Although the microorganisms degrading the polymer carbohydrates and lignin may be partly the same, the physiology of the degradation of celluloses and lignin is different as are the induced enzyme systems, as previously described. It would therefore be reasonable to use this physiological background for a definition of different steps in the decomposition process, based on substrate and nutrient availability. Raised nutrient levels, in particular N, P and S, that are normally the main limiting nutrients for microbial growth, stimulate microbial degradation of cellulose, hemicelluloses and many solubles.

The degradation of solubles and the early degradation of hemicelluloses and cellulose are rather rapid processes and the measured early-stage rates in a field experiment were at least twice as high as in the late stage. The relative degradation rates of the polymer carbohydrates are relatively high and range from 0.063 to 0.146% day^{-1} (Table 2.1). In the late stage, the rate of decomposition of the same components can range between 0.038 and 0.053% day $^{-1}$. The higher rate of disappearance of arabinan may be due to this hemicellulose being more easily hydrolyzed and/or less protected. Thus, the cellulose and hemicelluloses were degraded rather quickly until the unshielded portions were consumed. Investigations of the changing patterns of enzyme activities in decomposing litter also support this division in phases (Fig. 2.7). The cellulolytic enzyme appears relatively early, reaches a maximum and decreases before the peroxidase (part of the lignolytic system) appears.

The majority of studies on litter decomposition present results from the stage when the litter is recently shed, where normally positive relationships are seen between litter concentrations of N, P or S and factors such as the mass-loss rate or CO_2 release (Taylor et al. 1989; Berg et al. 1997). Also climatic factors have a strong influence on the turnover rate in newly shed litter (Jansson and Berg 1985; Berg et al. 1993a).

The simplified and incorrect picture is that climate regulates on a regional scale and substrate quality on a local one. This picture holds in a few cases but is far from general. We discuss the basic model with this reservation, while recognizing that local climate and nutrient availability appear to dominate the early stage of decomposition.

2.4.2 The late stage: lignin-regulated decomposition phase

In decomposing foliar litter there is an increase in concentration of the recalcitrant compound lignin and its recombination products that resemble lignin (Fig. 2.2).

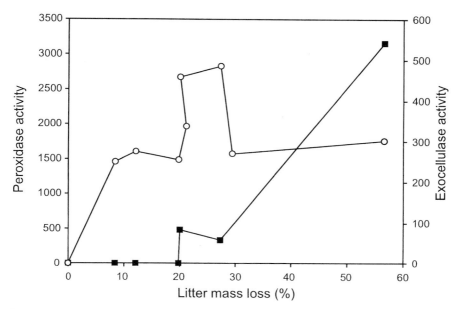

Fig. 2.7. The development of enzyme activity was followed during the decomposition of Scots pine needle litter. Cellulases (○) appear when the decomposition has proceeded for some time and a peroxidase (■) appears later, indicating the onset of lignin degradation. (C. McClaugherty unpubl.)

This may be explained by the fact that the lignin-degrading microorganisms, under the nutrient conditions which exist in most foliar litter types, grow more slowly than those degrading the polymer carbohydrates and that lignin as a chemical compound in foliar litter normally is degraded more slowly than many other components (cf. below). Eventually decomposition will reach a condition in which the litter contains primarily material that is rich in lignin and any remaining cellulose and hemicelluloses are enclosed and protected by lignin and newly formed humic compounds. The point at which this occurs varies among litter types and with environmental conditions. For Scots pine litter, Berg and Staaf (1980a) estimated this to be at around 30% mass loss. Under the conditions present in foliar litter (low levels of carbohydrate and relatively high N levels, see below), lignin is degraded slowly.

When it was found that a lignolytic fungi had an improved degradation of lignin in the presence of a more easily metabolized C source (e.g. cellulose) several attempts were made to explain differing lignin degradation rates using different indices. These indices were all based on relative amounts of cellulose, hemicellulose and lignin. Some investigators have used the lignin concentration itself (Fig. 2.8); others have created quotients between the lignin in litter and the holocellulose plus lignin parts (HLQ; Berg et al. 1984; McClaugherty and Berg 1987) or between lignin and lignocellulose (LCI; Aber et al. 1990). These developments still have not improved our understanding of the degradation rate of lignin and litter in late stage of decomposition, where the lignin degradation rate regulates the decomposition of the whole litter (Berg and Staaf 1980a; Berg et al. 1987).

The effect of lignin as a rate-retarding agent may be illustrated by a relationship between increasing lignin levels in litter and the decomposition rate of litter (Fig. 2.8A). Several scientists have observed this kind of relationship (Fogel and Cromack 1977; Meentemeyer 1978). A notable phenomenon (Meentemeyer 1978; Berg et al. 1993e) is that the influence of lignin on the decomposition rate is so strong that the effect of climate is not only suppressed but appears to disappear completely when lignin concentrations become high enough. This phenomenon has been studied further in a climatic transect with a range in annual average temperatures from ca. 0° to 8°C and extending from northern Scandinavia to northern Germany. In that study, using Scots pine needle litter incubated in litter bags in the field, it was found that in a warmer and wetter climate the effect related to lignin on litter mass-loss rate was stronger than under colder and drier conditions (Berg and Meentemeyer 2002; Johansson et al. 1995; Fig. 2.8B, see also Chap. 7).

Respiration rates from decomposing litter (and humus) often are used directly as measured, and in short-term experiments this means that the respiration rates are determined by those components that decompose the fastest. In an attempt to overcome this, Couteaux et al. (1998), using Scots pine needle litter, divided the litter into three compartments with components of different degradability and called them "labile", "intermediate", and "recalcitrant". Using different approaches, they determined the sizes of these pools and the rate constants for decomposition. The dominant pool was the recalcitrant one, making up between 79 and 85% of litter samples. The respiration rate for the recalcitrant fraction was lower than 0.0001% day^{-1} and for the labile one about 0.124% day^{-1} (Table 2.2). They found support for an early and a late decomposition phase with a transition at about 25% mass loss, but concluded that decomposition rates measured under laboratory conditions should be interpreted with care.

Effect of N on lignin degradation

As litter decomposes and the lignin level increases, the concentration of N also increases (Figs. 2.2, 5.2). This is a well-known and general phenomenon in decomposing litter. We emphasize this because N has an important role in the degradation of lignin and the formation of humus. This is discussed by Stevenson (1982) and described by among others, Nömmik and Vahtras (1982), and Eriksson et al. (1990).

There is no really clear border between the late stage and that which we call humus-near or the limit value stage of decomposition. Several of the functional properties, such as the effect(s) caused by N, appear to be common to both stages. This is an important observation that helps us to interpret and predict properties in humus.

The rate-retarding effect normally ascribed to the increasing lignin concentrations (Fogel and Cromack 1977; Berg and Lundmark 1987) is probably due more to the associated high concentration of N which has a suppressing effect on formation of ligninase and thus on lignin mass loss. Thus it is not the lignin concentration per se that is rate retarding, but the lignin in combination with N concentrations above a certain level. We may, as an extreme case, imagine that with low enough N levels and thus no suppressing effect, a higher degradation rate for lignin may result and probably no raised lignin concentration. In fact, there is evidence for this in wood decay because wood is extremely low in N. Based on litter bag studies in Black Hawk Island (Appendix III), wood chips of red maple and white pine lost more mass (88%) than any of ten other litter types including leaves, needles, bark and fine roots after 10 years of incubation (McClaugherty unpubl.).

Much remains unknown about the roles of lignin and N during this middle stage of decomposition. We do not know whether the rate-retarding effect ascribed to lignin, or the combination of lignin and N, in ageing litter is due to the level of N initially present in litter or if the availability of N from the outside is more important. The transport of N through fungal mycelia into the litter is one easily available source of N from outside the decomposing substrate (Berg 1988). Possibly both internal and external sources are influential, but this requires further investigation.

The question that arises is whether the declining rate of decomposition observed in the late stage of decay is due to lignin, N or a combination of the two. The suppressing effect of N on the degradation of lignin, as well as on the decomposition of whole litter, has been observed in studies with different levels of resolution, and is based on both organic-chemical observations (Nömmik and Vahtras 1982) and evidence from a microbiological-physiological level (reviews by Eriksson et al. 1990; Hatakka 2001). The effect of N has also been observed directly, producing different lignin-degradation rates in decomposition experiments (Berg and Ekbohm 1991).

There are two main paths recognized today which can influence degradation of lignin and its modified forms; one biological and the other chemical.

Biological mechanisms. Raised N levels may suppress the degradation of lignin (Keyser et al. 1978; Eriksson et al. 1990; Hatakka 2001) and consequently also the decomposition rate of litter (Berg et al. 1987; Berg and Ekbohm 1991; Berg and Matzner 1997). This simply means that the higher the level of available N, the stronger the repression of the formation of lignolytic enzymes in the population of lignin-degrading organisms (see also Chap. 3). The ability of several fungal species to degrade lignin was heavily suppressed when N was added to the culture medium at concentrations of 2.6-7.8 mM, corresponding to 0.0036-0.0109%. The level of N in solution in a pure fungal culture is not directly comparable to those in litter, where the N will be bound in different compounds and will be much less mobile than in solution. However, trends apparent in culture may stimulate speculation as to possible mechanisms in litter (Table 2.3). With an N concentration of 0.4% in our case study litter, N concentration is 100-fold greater than in the fungal culture system. In both cases the status of the N changes over the course of the experiment. In the liquid pure culture the N becomes bound in microbial biomass and thus less available. In the litter substrate there may be a mineralization and degradation of proteins, thus converting a fraction part of the bound N to more available N.

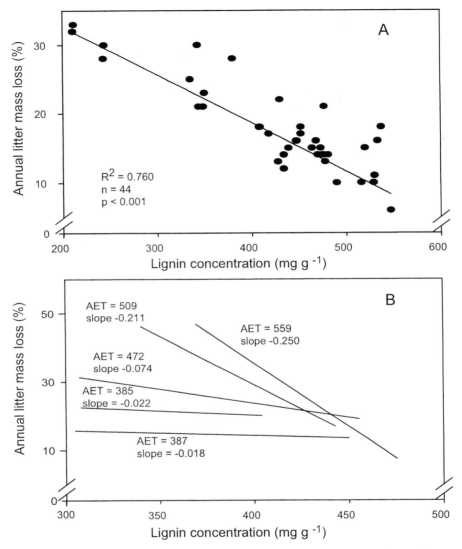

Fig. 2.8. Annual litter mass loss (%) as a function of initial lignin concentration at the start of each one-year period. Lignin in foliar litter is normally resistant to degradation, and an increasing lignin concentration suppresses the decomposition rate of the litter. The basic condition is that litter N levels are high enough to suppress the degradation of lignin. **A** The relationship indicated here, namely a decreasing rate with increasing lignin concentrations for one type of litter incubated at its own forest stand, has been observed by several scientists. **B** The same relationship under different climate conditions indicates that the rate-retarding effect of lignin is stronger in warmer and wetter climates. Redrawn from Berg et al. (1993e)

Table 2.2. Compartments of different stability in decomposing Scots pine needle litter and humus in a Scots pine forest. The sizes of the compartments were estimated and the rate constants were based on respiration measurements. Standard deviation in parentheses. (Couteaux et al. 1998)

Labile comp. [%]	K_L [% day^{-1}]	Intermediate comp. [%]	K_{IN} [% day^{-1}]	Recalcitrant comp. [%]	K_R [% day^{-1}]
Needle litter incubated in litter layer for 16 months					
4.09 (0.39)	0.124	17.01 (2.41)	0.0087	78.52	<0.0001
Brown needle litter from forest floor					
4.67 (0.61)	0.124	21.91 (1.54)	0.0087	74.93	<0.0001
H-layer particles < 2 mm					
0.00	0.124	9.80 (1.32)	0.0087	91.20	<0.0001

Table 2.3. Relative effects of N levels in liquid pure cultures on the wood degradation activity of some N-sensitive white-rot fungi. A comparison is made to the natural concentrations of N in Scots pine needle litter

Substrate	N addition	Effect
Wood[a]	None	No inhibition
Wood[a]	0.0036% (2.6 mM)	20-70% inhibition
Wood[a]	0.0109% (7.6 mM)	40-95% inhibition
Scots pine needles - natural levels[b]	0.4 - 1.2%	Delayed degradation

[a]Eriksson et al. (1990)
[b]Berg and Ekbohm (1991)

Chemical mechanisms. After litter has started to decompose, the lignin begins to incorporate N and condensation reactions take place which mark a first step in humification (Nömmik and Vahtras 1982; Stevenson 1982). These chemical transformations may result in structures that are not easily degradable by the soil microorganisms. These recalcitrant compounds, for example humins (Nömmik and Vahtras 1982; Bollag et al. 1983; Liu et al. 1985) may act as chemical barriers to the decomposing microorganisms. The concept was suggested by Mindermann (1968) and may explain why Staaf and Berg (1977) and others have found cellulose and hemicelluloses in well-developed A_{01} and A_{02} horizons.

The nature of the fixation of NH_3 to organic matter, both humus and decomposing litter, is unknown, although several mechanisms are plausible. One such mechanism involves phenolic groups that are formed during lignin degradation (Nömmik and Vahtras 1982). Methoxyl groups are removed from the lignin aromatic ring, forming phenolic groups which then may react with and bind NH_3. Fixation of NH_3 may also occur during degradation of both cellulose and hemicelluloses, where reducing ends are formed on the chains of carbohydrates (see above; Chap. 3), which can react with NH_3. A fixation mechanism has been suggested involving quinones, which are formed during lignin degradation, as side products from laccase or peroxidase acting on di-phenol rings (Nömmik and Vahtras 1982). These quinones, by reacting with ammonia via di-phenols, could be transferred to heterocyclic polymeric compounds (Fig. 2.9).

Thus phenolic groups, quinones, and also carbohydrates may react with NH_3, producing fixed N. In their review, Nömmik and Vahtras (1982) define the term "fixed NH_3" as the NH_3 that is retained by the soil organic matter after intensive extraction and leaching with either diluted mineral acid or neutral salt solutions.

The fixation process involves ammonia, not the ammonium ion, and therefore the reaction is faster at higher pH values. Also, amino acids react at higher rates at higher pH values for the same reason. Broadbent and Stevenson (1966) demonstrated that close to pH 9 the reaction was 10-20 times as fast as at pH 6 and below. The higher the level of N (or the higher the degree of humification) the lower was the NH_3-fixing capacity of the organic matter studied. For Scots pine needle litter, Axelsson and Berg (1988) largely confirmed these findings and estimated a fixation rate three times higher at pH 9 as compared to pH 5. They also found that Scots pine needle litter that had reached a higher accumulated mass loss and thus higher N concentrations, adsorbed/fixed less ^{15}N (Fig. 2.10).

In their review Nömmik and Vahtras (1982) point out that prolonged exposure of organic matter to NH_3 under aerobic conditions leads to degradation of humic-acid polymers by hydrolytic and oxidative processes, which results in the formation of low-molecular weight soluble compounds. The "chemical" mechanisms for slower decomposition that are suggested here are not yet confirmed experimentally.

2.4.4 Humus-near stage in litter decomposition - limit values

Literature that covers the functional transfer from partially decomposed litter to humus is rare. Nonetheless, moderately decomposed litter, the humus-near stages, and humus have properties in common, an example being concentrations of N and the rate-retardation that may be related to N concentration. Also analyses of decomposition rates (Couteaux et al. 1998) show similar values in the labile, intermediate and recalcitrant fractions, both for litter close to the limit value and in humus.

It has been possible to adapt mathematical functions to the accumulated mass loss in litter and with good statistical precision, estimate how far the decomposition process should continue. A number of such investigations have been carried out on our case-study litter. Within a given forest plot there is a certain repeatability of limit values. For example, Berg and Ekbohm (1993) found some homogeneity among limit values within groups of studies on decomposing litter of Scots pine and lodgepole pine. They also stated that the limit values of the two groups were significantly different. Berg et al. (1999b) published data for 11 studies on Scots pine litter decomposition in one forest system and found that the limit values ranged between 76.0 and 93.2%, giving an average of 84.7%. In other words, on average, 15.3% of the initial litter became stabilized as non-decomposing remains (Fig. 11.4). The SE of this estimate was 1.57.

The variation in concentrations of N and lignin in the needle litter collected in the different years (Johansson et al. 1995) may explain part of the variation in limit values (Table 2.4). For the same litter type, Couteaux et al. (1998) calculated a limit value of 96% and a mass loss rate measured as release of carbon dioxide that was less than $10^{-4}\%$ day^{-1} for the recalcitrant fraction (see above; Table 2.2).

It should be emphasized that the limit value pattern is by no means contradictory to those studies that describe two- or three-factorial models for decomposition, in which the different linear factors give different rates at different decomposition stages (see Chap. 10). Early work with such two-factorial models included work by Lousier and Parkinson (1976) and Berg et al. (1982b). Later, Couteaux et al. (1998) developed a three-factorial model that was used to estimate the decomposition rate at stages close to the limit value.

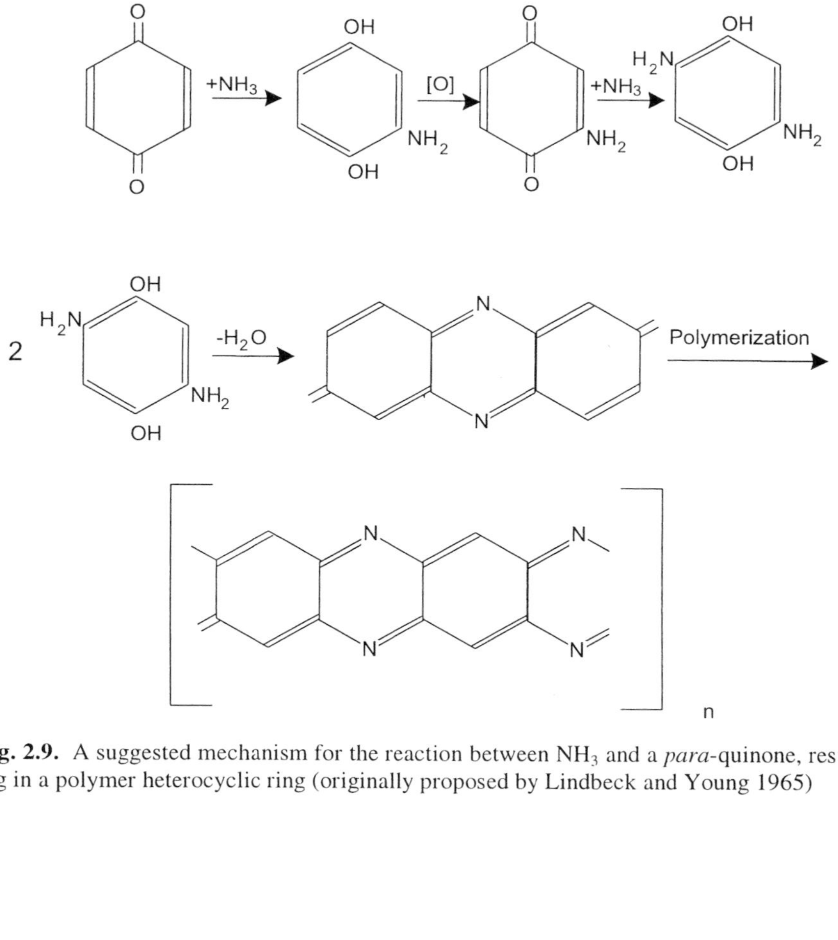

Fig. 2.9. A suggested mechanism for the reaction between NH_3 and a *para*-quinone, resulting in a polymer heterocyclic ring (originally proposed by Lindbeck and Young 1965)

Bosatta and Ågren (1985) proposed a conceptual mathematical model of decomposition. Their model was based on a continuous decline in substrate qualityduring decomposition. Depending on the relative rates of mass loss and quality decline, the model predicts different final outcomes for litter decay. One possible outcome is that all litter will disappear completely in a finite time; another outcome is that mass loss will eventually cease, leaving a residue for an indefinite time.

The role of animals in the decomposition of litter towards a limit value is unclear. Most of the studies that are used to estimate limit values have been carried out in forest soils containing relatively small numbers of animals that would attack litter. The existence of limit values and their levels in these studies, may in part be ascribed to the absence of soil animals. However, other studies have indicated that decomposition may have a limit value pattern in systems in which soil animals are found in higher numbers.

Table 2.4. Initial concentrations of N, Mn, water solubles and lignin in annually collected Scots pine needle. The Scots pine needle litter was incubated in the forest floor in its own system (SWECON site, Jädraås, Appendix III) and limit values for accumulated mass losses were calculated.

Initial concentrations [mg g^{-1}]				Limit value	
N	Mn	Wsol	Lignin	[%]	SE
4.0	n.d.	n.d.	267	93.2	11.70
3.8	1.55	92	223	86.6	3.02
4.0	n.d.	151	255	92.2	8.51
4.4	n.d.	n.d.	256	78.2	4.03
4.8	0.79	164	231	89.0	7.00
3.8	1.00	165	257	89.4	17.04
3.9	1.17	214	231	83.2	9.23
3.8	0.77	180	231	84.3	5.21
3.7	1.12	82	288	76.0	5.80
3.8	1.08	178	229	82.5	3.10
3.8	1.18	162	250	85.3	8.19
AVERAGES					
3.98	1.08	154	226.5	84.7	1.57

Wsol water-soluble substances, *n.d.* not determined

Although it is possible to estimate significant limit values for litter decomposition we do not conclude that such limit values necessarily indicate completely recalcitrant remains in the humus-near litter. The estimated values may illustrate a fraction that is stabilized and thus is decomposed at a very low rate or not at all. Even if this is the case, the phenomenon is no less interesting or useful, especially if we can connect this resistance or recalcitrance in litter to its properties, for example to the concentration of lignin or some nutrient, or to climate. Just the fact that allophanic (see Glossary) humus exists shows that an "eternal" storage is possible. Although allophanic organic material may be regarded as an extreme case, the level of stabilizing components (for example, aluminum and iron ions) necessary to stop the decomposition process is not known (Paul 1984). The fact that the use of limit values allows us to reconstruct a humus buildup over a period of 3000 years also indicates that the sources of error in undisturbed systems are probably minimal (Chap. 11).

What may cause the decomposition to cease?

Nitrogen. The influence of N on the lignin-degrading microbial population was discussed above. We described the rate-suppressing effect of N during the late and humus-near stages of decomposition. With N concentration increasing linearly with mass loss until the limit value is reached (Berg et al. 1999b), it would be reasonable to expect an increased effect of N at least until then. Although the chemical reaction between N compounds and lignin degradation products appears to be a slow process, the time required for our case-study litter to reach a mass-loss level within 5% of the limit value is of the magnitude of 8-10 years. Thus, even though the fixation process is very slow, especially at low pH values, it can become of significance over time. In a laboratory study, Axelsson and Berg (1988) found that newly shed Scots pine needle litter binds mineralized N in a pH dependent fashion, with 10 μg N g^{-1} litter bound in 24 h at pH 9.0, while at pH 5.0 the rate was about a third of that. The rate was probably dependent on the concentration of N and apparently the number of reactive sites, as well as the litter pH (Fig. 2.10, Table 2.5).

Table 2.5. The capacity of milled Scots pine needle litter to bind NH_3 in a solution of about pH 9.0. The incubation time was 24 h and the milled needle litter was extracted with 1M KCl. The rate at pH 5.0 was about a third of that at pH 9.0. (Axelsson and Berg 1988)

Mass loss [%]	Lignin conc [mg g^{-1}]	Total N conc [mg g^{-1}]	Excess of ^{15}N [10^3]	Bound ^{15}N [μg g^{-1}]
0	300	3.9	516	20.1
10.6	332	3.75	449	16.8
26.3	374	4.6	362	16.6
35.9	381	5.05	542	27.4
41.8	402	5.3	345	18.3
58.7	433	8.0	115	9.2
62.1	426	8.65	84	7.3
68.7	444	9.5	132	12.5

Manganese. Mn is essential for the activity of Mn peroxidase, a lignin-degrading enzyme, and enhances its production (Perez and Jeffries 1992). Mn is also involved in the regulation of other lignolytic enzymes, including laccase (Archibald and Roy 1992) and lignin peroxidase (Perez and Jeffries 1992). In a large number of laboratory studies it has been shown that this nutrient is essential for formation and activity of lignin-degrading enzymes and thus for degradation of lignin. Although the causal relationships are unclear, the concentration of Mn may influence the limit values of the litter (see Chap. 6).

When discussing the possible relationship of nutrients or heavy metals to limit values, their initial concentrations may be used as indices, because they are relatively conserved during decomposition. For N, the concentration increases linearly with accumulated mass loss (Chap. 7; Berg et al. 1999b) and the use of initial concentrations should thus not cause any problem when used as an index. For Mn such an increase does not normally take place and it remains to be determined how Mn levels could be used.

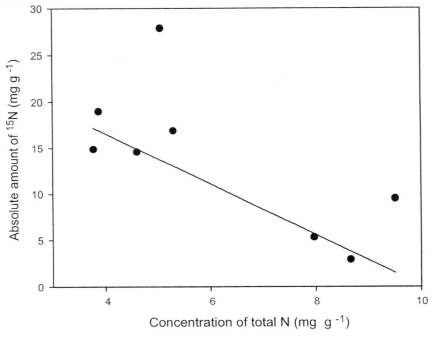

Fig. 2.10. Linear relationship between concentration of total N in Scots pine needle litter decomposed to different extents and absolute amount of ^{15}N (as NH_3) adsorbed g^{-1}.litter (Axelsson and Berg 1988)

3 Decomposer organisms

3.1 Introduction

The dominant primary decomposers in boreal and temperate forest soil systems are the microorganisms, encompassing both fungi and bacteria. Both these main groups of microorganisms can degrade cellulose, hemicellulose, and the different lignins. This chapter will emphasize the functional roles of organisms (e.g. cellulolytic and lignolytic) rather than their taxonomy. The concepts white-rot, brown-rot, and soft-rot and what they stand for functionally in terms of degradation processes will be presented.

Many microorganisms degrade cellulose and hemicellulose in nature. These organisms have in common the production of extracellular hydrolytic enzymes that are either bound onto the outside of the cell or released into the surrounding environment. Polymer carbohydrates may be degraded both aerobically and anaerobically, but a complete degradation of lignin (white-rot type) requires the action of aerobic organisms (fungi and/or aerobic bacteria). Partial lignin degradation (brown-rot type) may also be carried out by anaerobic bacteria, but is mainly found among fungi and aerobic bacteria.

We have used the functional concepts white-rot, brown-rot, and soft-rot as a basis for the discussion of degradation of litter. Although the terms originally seem to refer to visually different types of lignin degradation, it now appears that the degradation of cellulose and hemicellulose is also different among the groups (Worrall et al. 1997). The terms refer to the type of rot rather than to a group of organisms, but we have adopted the common use of the terms and refer to fungi when using the terms white-rot, brown-rot, and soft-rot. As regards degradation by bacteria, it is described and discussed as such.

The composition of the microbial population (e.g. cellulolytic *vs.* lignolytic) may vary with general properties of the soil/litter ecosystem, such as nutrient status and pH. A specific functional property that may discriminate among systems and populations is their sensitivity to N concentrations in litter and humus, which may be either stimulating or suppressing. Such sensitivity is not universal, but common in species of both white-rot and brown-rot organisms.

By tradition, soil animals such as collembola, mites and earthworms, among others, have been considered important for litter decomposition. Such groups have been ascribed different roles in decomposition, although those roles are not always clear. The decomposition by free-living microorganisms has also been considered important, but the relative influences of soil animals and soil microorganisms have not been apparent, thereby indirectly supporting studies of the more easily studied visible component, namely soil animals.

In recent decades, it has become increasingly clear that for some systems, at least boreal and temperate coniferous ones, the microbial component is of absolute dominance. For example, Persson et al. (1980) estimated that at least 95 % of the energy flow goes through the microbial population. The implications of this proportion are considerable and we could express this in a somewhat simplified way by stating that the decomposition of litter in a given system is determined by conditions and limitations that are valid for the microbial community of that system, which may be quite different from those of the soil faunal community.

Considering that the focus of this book is directed towards boreal systems, which have a decomposition process dominated by microorganisms, we have described both the microbial population (Sect. 3.2) and the microbial enzymatic degradation mechanisms (Sect. 3.3). This chapter will describe those properties of the organisms that are important in the degradation of cellulose, hemicellulose, and lignin. The main combined effects on the decomposition of whole litter are given in Chapters 2, 6 and 11.

Mycorrhizae have been found to turn into aggressive decomposers under certain circumstances and may decompose humus that has been considered as stabilized. Such a degradation can take place at a high rate. This phenomenon may be related to nutrient stress in growing trees. The role of mycorrhizae in decomposition in general is still under dispute and we present observations here without taking part in that dispute (Sect 3.4).

The ecology of decomposer communities can influence the pattern of decay. The changes in the community and its function during the decay process and the cyclic nature of the successional process in the soil will be addressed. The effects of moisture and temperature on the activity of the microbiological decomposition process are presented later, in Chapter 7.

3.2 General properties of a given microbial population

The two main groups of litter decomposers are bacteria (including the filamentous bacteria that earlier were called actinomycetes) and fungi: two groups that appear to include some of the same basic physiological properties when it comes to degradation of fresh litter polymers. Still, the fungi are generally considered the most important group, and we know more about their litter degrading properties and enzyme systems. Each of these two groups may be divided into functional subgroups with different properties, degrading different groups of chemical components. The taxonomy of both fungi and bacteria is complex and is beyond the scope of this book.

The bacteria include both aerobic and anaerobic organisms, which distinguishes them from the exclusively aerobic fungi. Both groups have organisms able to degrade all the main polymers: lignin, cellulose and the hemicelluloses. There are also organisms able to degrade woody tissue where all these components are

Table 3.1. Some general properties of the main groups bacteria and fungi

Property	Bacteria	Fungi
Mobility	+	+
Spore-forming ability	+	+
Can degrade cellulose/hemicellulose	+	+
Can degrade lignin completely	+	+
Can degrade lignin anaerobically[a]	+	-
Can degrade intact fiber walls	+	+
Species with N repression of the ligninase system	?[b]	+
Species without N repression of the ligninase system	?[b]	+

[a]Incomplete degradation to be compared to the brown-rot type
[b]Not known

combined into fibers. Complete degradation of lignin appears to be carried out by a small number of the fungi and aerobic bacteria. Some of the general properties of main groups of bacteria and fungi are collected in Table 3.1.

The biological diversity in the soil microbial community is high. The potential species diversity is evident just by comparing crude numbers of identifiable species. For just 1 m^2 of a given soil system we may estimate that for bacteria there may be 1000-5000 species and for fungi perhaps 100 dominant species.

The high density of bacteria in, for example, an organic soil, creates a high potential for invading a new substrate, such as newly shed litter. Estimates of 10^9 bacteria g^{-1} organic soil either active or in a resting stage are common when made by direct light microscopy counting. This figure is conservative since there are numerous bacteria that are simply too thin to be detected with light microscopy and have to be counted using electron microscopy. In the same soil, total mycelial lengths have been estimated to reach ca. 2000 km L^{-1} of humus, of which perhaps 10% would be live mycelium. Microorganisms will only be actively dividing and growing when environmental conditions are favourable. When the conditions cannot support growth the microorganisms will be in some kind of resistant, resting stage, or spore form. Wind and animals easily transport fungal and bacterial spores. This means that spores may be transplanted among ecosystems and that a given ecosystem may have a passive gene bank, able to quickly produce active microorganisms that can attack a particular litter type, possibly with new chemical components that are novel in a given environment.

The size of most microorganisms gives them access to different parts of the fiber and tissues that make up litter (Fig. 4.1). For a main part of the bacteria, the diameters range largely from 0.1 to 2 μm and for filamentous fungi from ca. 1 to less than 20 μm. The lengths of rod-shaped bacteria mainly range from ca. 1 to 10 μm, while those of the fungal mycelia are more undetermined.

Bacteria may be either immobile, or mobile, with one or more flagella, a whip-like structure. Fungal mycelia demonstrate mobility in another way, in that they simply grow in one direction and thus move their protoplasm, leaving an empty cell-wall structure behind.

The term "decomposer" microorganism is sometimes used for those microbes that decompose plant litter structures, sometimes for the larger group that decomposes organic matter, thus including the whole group of free-living heterotrophic microorganisms. Free-living in this context simply means those microorganisms that do not live in obligate symbiosis. Here, we will focus on what may be called primary litter decomposers, namely those that attack and degrade (at least in part) the polymer structures to carbon dioxide and/or small, partly degraded molecules. We discuss below the hypothesis that not only free-living microorganisms play a role in the turnover of organic matter but that mycorrhizal fungi may also be important.

3.3 The degradation of the main polymers in litter

3.3.1 Degradation of cellulose

Cellulose is degraded by numerous species of both bacteria and fungi. These organisms rely on extracellular enzymes that either are secreted into their immediate surroundings or are located on the cell surface. It is necessary that cellulose be degraded outside the microbial cell (Fig. 3.1), and that the insoluble macromolecules be degraded to monomers or oligomers of a few glucose units (Fig. 3.1), such as cellobiose, that can be taken into the cell and metabolized.

A common feature among all cellulose-degrading organisms is that they produce hydrolytic, extracellular enzymes that attack the cellulose polymer. Part of the cellulose in the plant fiber is arranged in a crystalline form that makes it harder to attack (see Chap. 4) and relatively few cellulolytic organisms have the necessary complete set of enzymes to degrade this structure. Many organisms are able to degrade the more amorphous kind of cellulose (see Eriksson et al. 1990).

The most studied group of cellulose-degrading organisms is the fungi and no less than 74 species (Eriksson et al. 1990) have been studied in some detail. The traditional division of wood-degrading fungi into three main groups, white-rot, brown-rot, and soft-rot fungi, relates primarily to their mode of lignin degradation, but these groups also differ in the way they degrade polymer carbohydrates.

Perhaps the most studied wood decay fungus is the white-rot basidiomycete *Phanerochaete chrysosporium* Burdsall (previously called *Sporotrichum pulvurolentum* Novabranova). Much of what is known about the decay of lignocellulosic materials in nature is based on studies of this fungus (Ander and Eriksson 1977; Higuchi 1993; Tien and Kirk 1984). Three main hydrolytic enzymes are involved in cellulose degradation; one type of enzyme (endo-1, 4-β-glucanase) covers the cellulose chain and splits the glucosidic linkages in a random fashion (Fig. 3.1) producing oligosaccharide units of different lengths that may still be attached to the micro-fibril structure. Another enzyme, an exo-1, 4-β-glucanase, splits off either cellobiose or glucose from the non-reducing end of the cellulose chain. Finally, 1,4-β-glucosidase hydrolyzes cellobiose and other water-soluble oligosaccharides, such as triose and tetraose, to glucose. One important aspect of this enzyme system is that the different enzymes with different specificities (the endo-

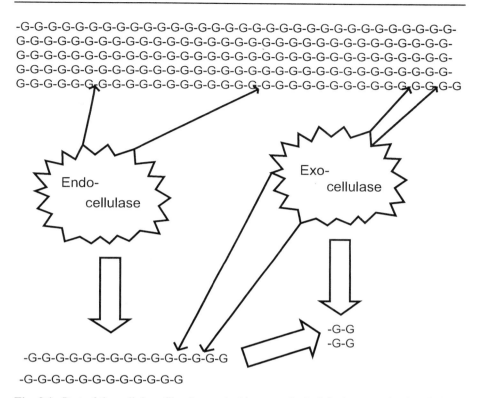

Fig. 3.1. Part of the cellulose fiber is attacked by an endo-1, 4-β-glucanase (endocellulase) breaking the chains and splitting off oligosaccharides in a random manner, including soluble shorter chains with, for example, 3 to 5 glucose units. An exo-1, 4-β-glucanase (exocellulase) splits off cellobiose units from the non-reducing end of the carbohydrate chains. *G* stands for a glucose unit

and exo-glucanases) exert a synergistic action that enables them to degrade both crystalline and amorphous cellulose.

In addition to hydrolytic enzymes, some cellulolytic organisms produce cellobiose dehydrogenase. This enzyme is found in a variety of fungi and appears to have roles in both lignin and cellulose degradation. There was much confusion in the early literature about this enzyme, but recent work has resulted in a renaming of both cellobiose oxidase and cellobiose:quinone oxidoreductase to cellobiose dehydrogenase (Cameron and Aust 2001).

The soft-rot fungi as a group appear to have a cellulose-degrading system similar to that of the white-rots. However, in contrast to white-rot fungi, brown-rots have not been found to have the synergistic enzymes that are found in white-rots and they appear not to have the exoglucanase mentioned above. However, Highley (1988) found several species of brown-rotters that were able to solubilize microcrystalline cellulose. Thus, the generally held conclusion that these fungi merely seem to depolymerize cellulose without producing soluble monomers or dimers may not be entirely correct. Still, no other enzyme has been found to substitute for the missing exoglucanase that splits off soluble units. This has led Eriksson et al. (1990) to conclude that there may be a non-enzymatic mechanism involved. An observation that hemicellulose is virtually absent in wood decayed by brown-rots, suggests that brown-rot fungi may degrade hemicelluloses. Although the mechanisms for degradation of cellulose are far from clear, work on a basidiomycete (Wolter et al. 1980) suggested that at least for some species, a less specific or multifunctional enzyme that could degrade several different polysaccharides was active, an observation that suggests that this enzyme also has an effect on cellulose.

The ability to degrade crystalline cellulose is also found in many bacteria. Detailed studies on *Clostridium cellulolyticum* show that the organism produces at least six different cellulases, each with slightly different structural and catalytic properties. The cellulases, along with xylanases, are held together in a large structure, the cellulosome, by a scaffolding protein (Bélaich et al. 1997), much as was envisioned by Eriksson et al. (1990). In the anaerobic bacterium *Clostridium thermocellum*, a multi-component complex of cellulolytic enzymes was named "cellulosome" in the very early work of Viljoen et al. (1926). A close contact between the cellulose substrate and the organism often appears to be necessary.

The degradation of cellulose by bacteria has been suggested to be hydrolytic; although the mechanisms seem to be different from those found in fungi. For bacteria, the cellulolytic enzymes are arranged in clusters and act in a combined way as described above. There are a few other groups of cellulolytic bacteria that have been studied including: *Cytophaga*, *Cellulomonas*, *Pseudomonas* and *Cellvibrio*. It appears that these have their cellulolytic enzymes bound to the cell wall and therefore a close contact is needed between the cell and the substrate (Berg et al. 1972; Eriksson et al. 1990). This property seems today to be widely recognized (Wiegel and Dykstra 1984).

Actinomycetes, in contrast to some other bacterial groups, appear to degrade cellulose in a manner similar to that of the fungi and can also degrade the crystalline form. Several strains have the ability to degrade the lignocellulose complex. The fungal model for enzymatic attack on the cellulose molecule, namely that an endo- and an exocellulase act synergistically, appears to be valid for actinomycetes, supporting their similarity to white-rot and soft-rot fungi.

The synthesis of cellulases is induced by cellulose, cellobiose, sophorose, and lactose. The presence of cellulose appears to be the best induction agent. On the other hand, the presence of glucose seems to repress the synthesis of the cellulase system. As cellulose is a large and non-soluble molecule, it cannot be absorbed into the microbial cells and exert an inducing effect. Today, the accepted theory is that the organisms have a constant, basic level of cellulase on their surface. Upon contact with cellulose, low amounts of inducing substances are released from the cellulose, enter the microbial cell, and induce cellulase formation. It is likely that both the type of compound, for example cellobiose or cellotriose, and a low intracellular concentration of these compounds influence the synthesis of cellulase. There are also theories that metabolic transfer products of glucosyl are active as inducing agents, one of these being sophorose (Eriksson et al. 1990).

3.3.2 Degradation of hemicelluloses

In wood, the total concentration of hemicelluloses usually ranges from 20 to 30% (Chap. 4). There are clear differences in the composition and structure of hemicelluloses in softwood as compared to hardwood litters. The composition of hemicelluloses is clearly different between hardwoods and softwoods (Table 4.1). The hemicelluloses are composed of both linear and branched heteropolymers of D-xylose, L-arabinose, D-mannose, D-glucose, D-galactose, and D-glucuronic acid. These individual sugars may be methylated or acetylated, and most hemicellulose chains contain between two and six different kinds of sugars. Hemicelluloses from hardwoods have average degrees of polymerization in the range of 150-200 units and most hemicelluloses are based on the 1,4-ß-linkage of their main sugars.

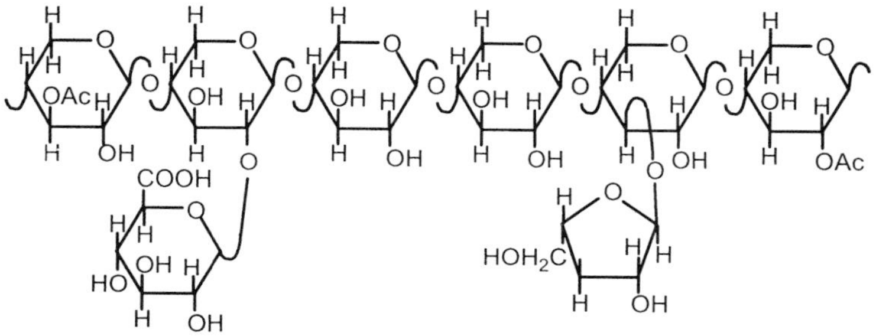

Fig. 3.2. An example of a fragment of a xylan molecule. The backbone of the molecule is made up of xylan units of which part are acetylated (Ac) and part not. The branches in this case are composed of glucose (left) and arabinose (right) units. The main enzyme attacking the unbranched part of the chain would be an endo-1, 4-β-xylanase, producing oligomers of different lengths. β-xylosidases split the oligomers into simple xylose units. Other enzymes are necessary to split off the side chains as well as for example, the acetyl substituent. (Eriksson et al. 1990)

Degradation of hemicelluloses requires more complex enzyme systems than are needed for the hydrolysis of cellulose. For example, xylan-based hemicellulose contains both 1,4-ß-linkages and branched heteropolysaccharides, which require a complex set of enzymes for degradation (Dekker 1985). Figure 3.2 shows the possible structure of a xylan-dominated molecule. The xylan backbone is made up of both acetylated and nonacetylated sugar units. On the branches, there are units of glucose and arabinose. The degradation of such a molecule requires the concerted action of several different hydrolytic enzymes (Eriksson et al. 1990).

3.3.3 Degradation of lignin

Lignin degradation is regarded as a process that differs among the three general groups of decomposers: white-rot, soft-rot, and brown-rot fungi. Although the names are old and refer to characteristics easily seen by the eye, there are also functional differences in the degradation mechanisms, motivating the continued use of the terminology. The names are used in connection with fungi although bacteria are also lignin degraders.

The number of different enzymatic mechanisms of lignin degradation with which organisms operate appears to be large, and only a few are well described. In fact, today it appears that only one mechanism of lignin degradation is well described, namely that for *Phanerochaete chrysosporium*, a white-rot fungus. Some characteristics for each of the groups are given below, starting with white-rots since these are not only the most studied ones, but also probably the strongest lignin degraders known.

Lignin degradation by white-rot fungi

White-rot fungi possess the ability to completely mineralize lignin to CO_2 and H_2O. The result, for wood, is that the entire lignocellulosic complex is degraded more or less simultaneously. A large group of the white-rots may even degrade lignin preferentially to cellulose (Hatakka 2001).

The attack on lignin structure has long been considered to start with a removal of the methoxyl group (Figs. 3.3 and 3.4). Newer research has shown that a combination of hydroxylation and demethylation is followed by an oxidative attack on the aromatic ring (Eriksson et al. 1990). The cleavage of the aromatic ring (Fig. 3.4) is an oxygen-demanding step and the data in Table 3.2 illustrate the importance of the presence of O_2.

The lignolytic enzyme system of our example fungus (*P. chrysosporium*) is synthesized as part of several physiological events that appear to be triggered by N starvation. As described by Kirk (1980) a whole set of enzymes are synthesized under conditions of N starvation (see below). Almost all white-rot fungi produce Mn peroxidase, a fact that may create an ecological niche, based on Mn as a limiting nutrient.

Although we may know more about the lignolytic system of *P. chrysosporium* than those of other white-rots, it appears that the lignolytic systems are species specific and it has been suggested that they depend on the ecological niche of the fungus in question (Hatakka 2001). For example, the white-rot *Ganoderma lucidum* produces Mn peroxidase in a medium with poplar wood but not in one with pine wood (D'Souza et al. 1999). An observation like this latter one may support the finding that white rot fungi are more commonly found on angiosperm than on gymnosperm wood (Gilbertson 1980).

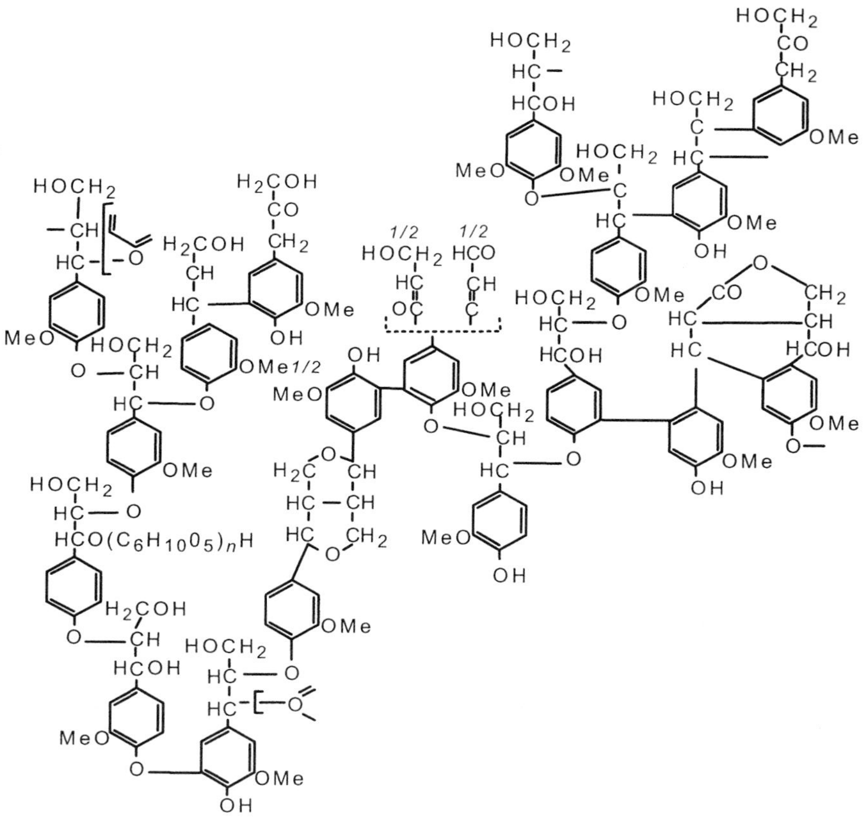

Fig. 3.3. Part of a lignin molecule from spruce

Lignin degradation by brown-rot fungi

Brown-rot fungi decompose mainly the cellulose and hemicellulose components in wood and have the ability to significantly modify the lignin molecule, but are not able to completely mineralize the compound (Eriksson et al. 1990). They allow for the degradation of cellulose with a relatively small loss of lignin mass.

Brown-rot fungi are considered to have similarities in degradation mechanisms to white rot fungi. In both cases, the formation of hydroxyl radicals that attack wood components is important and high oxygen tensions support the degradation (Hatakka 2001). It has been assumed that all brown-rot fungi use the same mechanism for wood decay. However, newer research has indicated, that in a parallel with white-rots, brown-rot fungi appear to have different mechanisms. The initiation of the degradation of both lignin and cellulose appears to be by diffusible small molecules that can penetrate the cell wall. In contrast to white-rots, only one brown-rot has been found to produce Mn peroxidase.

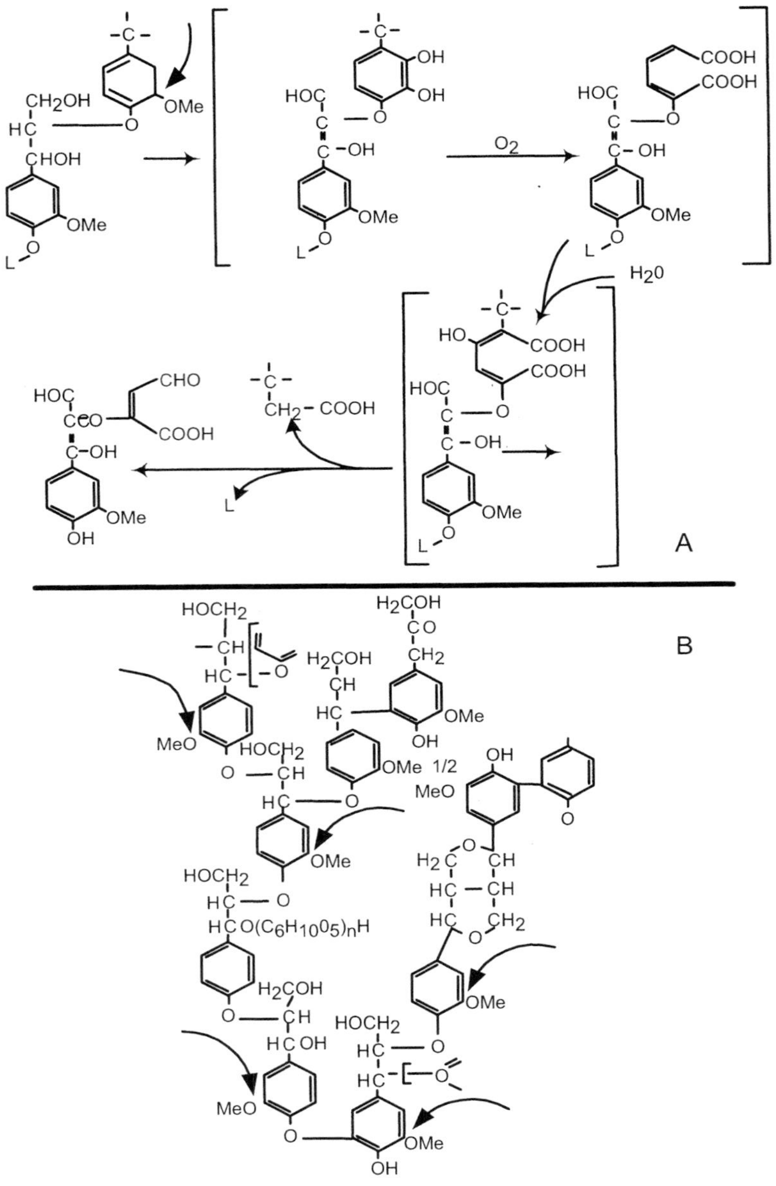

Fig. 3.4. Part of a lignin molecule of spruce during degradation. **A** Under degradation by white-rot a demethoxylation and hydroxylation are followed by an oxidative step leading to ring cleavage (from Kirk 1984). **B** The same molecule under attack by brown-rot fungi, resulting in just a demethylation where the methoxy groups (MeO) are replaced by an OH group

Table 3.2. Degradation of aspen wood lignin by different white-rot fungi in the presence of air or pure oxygen. Determinations were made as $^{14}CO_2$ evolution and as Klason lignin. (Reid and Seifert 1982)

Fungal species	$^{14}CO_2$ evolution [%]		Klason lignin loss [%]	
	Air	O_2	Air	O_2
Phanaerochaete chrysosporium	10.8	35.2	13	40
Coriolus versicolor	14.6	35.5	24	46
Gloeoporus dichrou	9.7	18.1	22	24
Polyporus brumalis	16.6	33.0	19	33
Merulius tremellosus	14.0	22.3	30	40
Pychnoporus cinnabarinus	13.6	22.6	18	37
Lentinus edodes	9.7	18.0	18	41
Bondarzewia berkeleyi	9.0	13.8	25	27
Pleorotus ostreatus	11.7	11.6	17	17
Grifola frondoza	9.2	10.6	8	15

The radicals formed by brown-rot fungi can remove methoxyl groups from lignin and produce methanol, leaving residues of mainly modified lignin (Eriksson et al. 1990). Relative to native lignin, brown-rotted lignins are structurally modified and have a decreased number of methoxyl groups (Fig. 3.4) and an increase in phenolic hydroxyl groups (Crawford 1981). Brown-rotted lignin is more reactive than native lignin due to the increased content of phenolic hydroxyl groups. Carbonyl and carboxyl groups are also formed (Jin et al. 1990)

Lignin degradation by soft-rot fungi

The traditional view has been that soft-rot fungi do not degrade lignin, but act to soften wood by breaking down the middle lamella of the cell wall (Fig. 4.1). Most soft-rot fungi are ascomycetes and deuteromycetes and are most active in moist wood. Crawford (1981) reviews a number of studies in which purported soft-rot fungi were able to decrease the lignin content of decaying wood.

Today it has been well confirmed that soft-rot fungi do degrade lignin: in laboratory experiments up to 44% was degraded at a wood mass loss of 77% (Nilsson et al. 1989). In general, they are considered to degrade lignin to some extent.

Evidence from a study on the fungus *Daldinia concentrica* may explain why these fungi preferentially degrade hardwoods. This fungus degraded birch wood efficiently but not that of pine (Nilsson et al. 1989). The lignolytic peroxididases of soft-rot fungi do not have the potential to oxidize the softwood lignin which have a high level of guaiacyl units. In contrast, soft-rot fungi readily oxidize the syringyl lignin in hardwoods (Nilsson et al. 1989).

Enzymes directly affected by Mn concentration in the substrate

Manganese-peroxidase belongs in a group of enzymes that are classified as phenoloxidases. Manganese is essential for the activity of the lignin-degrading Mn-peroxidase (Perez and Jeffries 1992). Although not much was published on this enzyme before 1983, Lindeberg (1944) discovered in the 1930s that *Marasmius* spp. were dependent on Mn for their growth and that a low level of Mn in a substrate hampered the degradation of lignin. This finding was not pursued and it was not until the 1980s that additional detailed studies followed.

Manganese is also involved in the regulation of other lignolytic enzymes, including laccase (Archibald and Roy 1992) and lignin peroxidase (Perez and Jeffries 1992). The role of Mn-peroxidase in lignin degradation is not clear although one of its roles may be to form H_2O_2. The enzyme itself shows no affinity for non-phenolic compounds, which on the other hand are readily attacked by ligninase. Blanchette (1984) found that Mn often accumulates as MnO_2 in wood attacked by white-rots, which suggests that Mn-peroxidases are important for the degradation of lignin. It has also been found that MnO_2 stabilizes lignin peroxidase.

Effect of N starvation on lignin metabolism

Lignin degradation may be repressed by high N levels in the substrate, an effect seen mainly in white-rot fungi but also in brown-rots and soft-rots. As mentioned above, Kirk (1980) described a set of effects for *P. chrysosporium* that were regulated by N starvation. A drastic effect on lignin degradation was seen when the N concentration in the culture medium was increased from 2.6 to 5.6 mM (Keyser et al. 1978), namely, the lignolytic activity (measured as transformation of ^{14}C-lignin to [14]CO_2) was repressed by 83%. The same property has since been described for several fungal species in laboratory experiments with pure cultures, although the levels of N and the magnitude of the effect vary. For three species (*Phlebia brevispora, Coriolus versicolor*, and *Pholiota mutabilis*), there were effects at 7.8 and 34 mM N in the culture, but not at 2.6 mM N. The magnitude of the effect varied from an almost complete repression in *P. chrysosporium,* to about approximately a 50% repression in *P. mutabilis*. When using ^{14}C-labeled lignin from red maple wood, there was a clear effect of 20 mM N. There are also several fungi that are not sensitive to N. For example, a white-rot strain isolated from an N rich environment (cattle dung) showed no sensitivity to raised N concentrations. Table 3.3 lists a number of species investigated for this property.

The results suggest that repression of lignin degradation by N is common but not always the rule. The addition of N to fungal cultures may in certain cases even increase their ability to utilize lignin. We would expect that such fungi, and tolerant fungi in general, would be found in environments with high N concentrations, as in the example given above with cattle dung whereas most white-rot fungi that grow in and on wood are adapted to low N concentrations. Many of the fungi that have been studied were isolated from wood, and the low N content in

Table 3.3. Some fungal species for which raised N concentrations have, or alternatively, have not elicited a repressing effect on lignin degradation.

Species	Comments	Reference
Sensitive to N		
Phanerochaete chrysosporium	Isolated from wood	Keyser et al. (1978)
		Eriksson et al. (1990)
Phlebia brevispora		Leatham and Kirk (1983)
Coriolus versicolor		Leatham and Kirk (1983)
Heterobasidion annosum	Some repression	Bono et al. (1984)
Not sensitive to N		
Pleurotus ostreatus		Freer and Detroy (1982)
Lentinus edodes		Leatham and Kirk (1983)
NRRL 6464 not identified	Isolated from cattle dung	Freer and Detroy (1982)

wood (with C-to-N ratios in the range from 350 to 500) may explain the generally strong influence of increased N levels.

Effect of the C source on lignin degradation

It appears that the presence of a carbon source other than lignin stimulates lignin degradation in several white-rot species including *P. chrysosporium, C. versicolor, Coriolus hirsutus, Polyporus* spp. and *Lentinus edodes.* One observation was that cellulose had a stronger stimulating effect on lignin degradation than glucose, an observation that was ascribed to its *lower* availability, thus an influence of catabolite repression could be expected (cf. Sect. 3.3.1). The major organic components in litter are normally the insoluble ones such as lignin, cellulose and hemicelluloses. The latter two normally supply the lignin-degrading organisms with alternative carbon sources.

3.4 Degradation of fibers

3.4.1 Bacteria

Though bacteria have long been known to be involved in litter decomposition, they have received far less study than have fungi. In most cases, bacteria coexist with fungi, particularly basidiomycetes and yeasts, and their presence has been shown to double the rate of fungal growth on wood and increase the overall rate of decay (Blanchette and Shaw 1978). Although it was once thought that bacteria were not capable of degrading lignified cell walls without some type of pretreatment, a variety of fiber degrading bacteria have now been identified. Three types of bacterial degradation are recognized, based on the manner in which they degrade the cell walls of the substrate: tunneling, erosion and cavitation (Blanchette 1995). Bacterial decomposition seems to be more common in situations where fungi are under stress. Bacteria have also been found to degrade substrates, especially wood, that are resistant to fungal decay (Singh et al. 1987).

3.4.2 Soft-rot

Soft-rots generally occur under conditions that are not favorable for basidiomycetes. However, a key for good growth of soft-rots is a high availability of nutrients. It is also generally held that soft-rots require moist conditions, though this requirement may not be different from that of basidiomycetes (Worrall et al. 1991). Two forms of soft-rots are identified based on the morphology of the degradation they cause (Blanchette 1995). Type I causes the formation of cavities in the secondary wall and is most commonly found in conifers, where lignin-like materials accumulate on the edge of the cavities. Type II causes cell wall erosion, but unlike white-rot, it does not degrade the middle lamella (Fig. 4.1). It is possible that the middle lamella is resistant because it contains more guaiacyl-propane units. Type II is more common in angiosperms.

3.4.3 Brown-rot

Brown-rot fungi have the ability to degrade holocellulose in plant cell walls without first removing lignin. Brown-rots apparently begin their attack on fibers by degrading the hemicellulose matrix because xylans begin disappearing before cellulose (Highley 1987). They do this by first causing a rapid decrease in the degree of polymerization of the holocellulose polymers. The decomposition occurs in a diffuse manner and, in wood, with a rapid loss of strength. These two factors suggest that agents smaller than enzymes are involved (Green and Highley 1997). This initial degradation is generally accompanied by relatively little mass loss.

When attacking fibers, brown-rot fungi appear to attack the S2 layer first, leaving the S3 layer until later (Fig. 4.1; Highley et al. 1985). The reason for this is not known, but Hirano et al. (1997) offer a proposed mechanism that agrees with the observations. They suggest that the brown-rot fungus grows into the cell lumen and releases a low molecular weight substance (1 to 5 kDa) that diffuses into the S2 layer. Fe (III) is then reduced to Fe (II) and chelates it. The newly formed complex with the Fe (II) catalyzes a redox reaction that produces hydroxyl radicals. These hydroxyl radicals are able to cut canals through the S3 layer large enough for cellulases to penetrate. Clearly, more work is needed to validate this mechanism and to identify the unknown substances required for its operation.

3.4.4 White-rot

White-rot fungi carry out two different types of fiber degradation: simultaneous rot and selective lignin degradation. Some species can carry out both types (Blanchette 1991). In simultaneous rot, the fungi are able to either erode the cell wall adjacent to the hyphae, creating erosion channels, or they generally erode the lumen surface, resulting in an overall thinning of the cell wall. In addition, the hyphae move from cell to cell through pits or by boring through the wall. The other type, selective delignification, often results in cell separation as well as overall thinning of the cell walls. Anagnost (1998) provides numerous photomicrographs that illustrate the various types of decay.

White-rots sometimes seem to have a delay or a lag time of relatively slow mass loss before a period of more rapid mass loss (Fig. 9.1). Blanchette et al. (1997) used a novel biotechnological approach to demonstrate why this might occur. They incubated loblolly pine wood with a white-rot fungus, *Ceriporiopsis subvermispora*. They then placed the wood, in various stages of decay, in solutions containing proteins of known size. Using immunocytochemical techniques, they were able to show that proteins of the size of cellulases and lignin degrading enzymes could not freely pass through the wood until later stages of decay. After cell walls had been thinned enough to increase their porosity, it was possible for extracellular enzymes to move freely from lumen to lumen, thus initiating the stage characterized by a higher rate of mass loss.

3.5 Mycorrhizae

In undisturbed soil systems, there also appear to be mechanisms that can change the composition of the microflora in ways that enhance its ability to degrade the otherwise stable humus. Hintikka and Näykki (1967) gave a good description of the mycorrhizal basidiomycete *Hednellum ferrugineum* and its effects on the humus layer. The development of thick mycelial mats under the mor layers was described, as well as bursts of soil respiratory activity, followed by a large decrease in the amount of humus in the FH layer. The effect was observed on dry, sandy, nutrient-poor sediment and till soils and could be attributed to plant growth. It appears to be a powerful mechanism driving humus decomposition. Unestam (1991) discussed this effect for certain other mycorrhizal fungi. Further, Griffiths et al. (1990) studied the effects of the ectomycorrhizal fungus *Hysterangium setchelli* on respiration in humus under Douglas-fir and identified patches with very high respiratory activity.

3.6 Ecological aspects

The composition of the microbial community that invades litter depends on the properties of the litter that falls onto the soil system and the changes in those properties over time. The community of decomposers undergoes many of the same ecological processes that act on communities of primary producers. These processes include succession and competition, while the pathway of decay may be influenced by modifications in these processes.

Microbial succession, the change in community composition over time, occurs as the quality of the decomposing substrate changes, but it also occurs because different organisms invade substrates at different rates. Griffith and Boddy (1990) followed the development of the fungal community on common ash, common oak and European beech twigs. The primary colonizers included endophytes that were present on the twigs while they were still alive. Secondary invaders were not endophytic and did not show up in appreciable numbers until about 11 months after twig death. They identified a third type of colonizer, the superficial, that appeared on the surface rather early into decay but was not present on the living twig. This pattern is probably similar for all litter types, though of course the species and timing may differ. For example, spruce needles can persist on the twigs for some time after death and decomposition can begin then. However, when the needles ultimately fall to the forest floor, the changing environmental conditions and the availability of a rich variety of inocula result in a change in the microbial community.

In addition to the change in microbial community that occurs along with decay, there are seasonal changes in the microbial community reflecting temperature and moisture. For example, Kayang (2001) followed fungi, bacteria, and selected enzyme activities in freshly fallen leaves on Nepalese alder in India. The climate was described as subtropical monsoon. The dry season occurred from November through March, with frosts during December and January. The fungal and bacterial propagule numbers varied by a factor of nearly five between winter and summer. Enzyme activities (invertase, cellulase and amylase) reached their peaks before the peak of microbial numbers, between April and June, and then fell slowly. The sequence of peaks suggests a succession of enzyme activities, with invertase, an enzyme involved in sucrose metabolism, peaking first. Amylase, which catalyzes the hydrolysis of starch, and cellulase appear later.

Examining activities of cellulases and cellobiose dehydrogenase on leaf litter in laboratory microcosms, Linkins et al. (1990) observed similar patterns for three different litter species. The species, flowering dogwood, red maple and chestnut oak, differed in lignin contents and decay rates. However, all three species exhibited an increase in cellulase activity that reached a peak at the same time that cellulose disappearance rate was at its maximum. Cellulase activity then began to decline and cellobiose dehydrogenase activity began to increase (Fig. 3.5).

As enzyme activities are changing, so are the fungal communities. Osono and Takeda (2001) studied the fungal populations on Japanese beech leaves as they

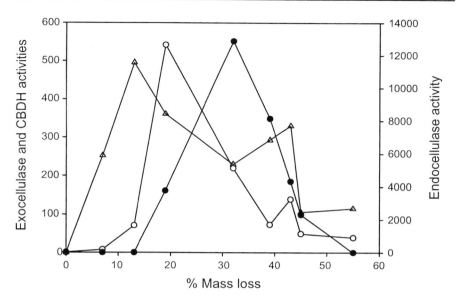

Fig. 3.5. Activities of exocellulase (Δ), endocellulase (○), and cellobiose dehydrogenase (CBDH) (●) during chestnut oak leaf litter decay in microcosms (Linkins et al. 1990)

decomposed in a cool temperate deciduous forest. Total and living fungal bio-mass, estimated using a modified Jones and Mollison (1948) technique (Ono 1998), increased during the first year of decay and then fluctuated for the remainder of the study period. The percentage of fungi that were basidiomycetes increased for the first 21 months of the study, reaching a maximum of 25 to 35% of the total living fungal biomass. They noted that the relative abundance of basidiomycetes was linearly and negatively related to the lignocellulose index (Chap. 2), an index of litter quality equal to the fraction of holocellulose in the lignocellulose. They identified over 100 fungal taxa on the beech leaves during their study and distinguished three groups: an early-appearing group, a late-appearing group, and a constantly appearing group. The early-appearing fungi were present during the period of net nutrient immobilization and the late-appearing fungi increased in number as the litter moved into the phase of net mineralization.

Decomposer populations may work synergistically or in competition. Competition is visible in decaying logs where discrete zones of decay caused by different organisms can be easily discerned. In some cases, the organisms define their boundaries with black zone-lines. The interspecies dynamic can change as decomposition proceeds. For example, Bengtsson (1992) found a synergism with no evidence of competition between fungi and bacteria on European beech leaves during their first year of decay in stream microcosms. In comparison, Møller at al. (1999) found clear evidence of competition between fungi and bacteria on 1-year-old beech leaf litter, also in a microcosm study. This difference probably relates to the age, and hence the state of decomposition and the quality of the litter.

Though there are not many studies on this phenomenon, it is possible that as litter quality decreases, the competition for the remaining resource becomes more intense.

As decomposition proceeds, the microorganisms themselves can become important substrates for the microbial community. Some fungi, including wood decay fungi, are able to use the cell walls of other fungi or bacteria, presumably as an N source. Some bacteria are able to degrade hyphal walls (Tsuneda and Thorn 1995).

There are many interactions among the organisms involved in decomposition and these interactions change over time. These complex, dynamic systems are not easily described. However, this natural complexity does have implications for the interpretation of pure culture and microcosm studies. Such studies are often the only way to control variability enough to ask a precise question. On the other hand, the behavior of a single, isolated species or of a simple community in a mesocosm may not reflect its behavior in the more complex natural environment.

4 Initial litter chemical composition

4.1 Introduction

In forested ecosystems, litter fall is the largest source of organic material and nutrients for the humus layer. The quality and quantity of litter fall influences nature of the microbial community, including its size, composition, function and physiological properties. The composition of the microbial community may, in turn, influence the course of decomposition and the chemical changes in the litter during decomposition. With knowledge about the initial chemical composition of litter and the chemical changes during decomposition, it is possible to predict how mass-loss rates will change even in late decomposition stages. With a close connection between the chemical composition of newly shed litter and the relative amount of recalcitrant residual litter (Chap. 11), we see a direct connection between litter chemical composition and rate of humus (soil organic matter) buildup.

Plants shed not only foliar litter, but, with a tree as an example - twigs, branches, bark, roots, flowers, and occasionally cones. Structures such as cones are often quantitatively important and may sometimes exceed foliar litter as the largest component. Several litter types are not "recently dead" but are recognized as litter after they have been shed and started to decompose and their chemical composition has begun to change. This applies, for example, to twigs and branches that stay on the trees as "attached dead" and often start decomposing before they fall to the ground and are recognized as litter. Roots die and are "shed" differentially based on their size and function, and dead roots may remain attached to their parent tree for extended periods.

The combinations of main chemical components in live vascular plants have general similarities among species and genera. Quantitatively dominant groups of polymer carbohydrates and lignin are ubiquitous. However, their proportions vary and minor structural differences occur among species.

Similarly, the same plant nutrients are found in the different plant materials and in the litter, even if in different proportions. All plant litter contains essential nutrients such as N, P, S, K, Ca, Mg, Mn and Fe, but the concentrations vary with the litter species. For example, leaf litter of the N_2 fixing genus alder *(Alnus)* has very high levels of N (often above 3%); in contrast, pine needle litter is more N-poor (often below 0.4%). Species is thus one dominant factor when it comes to determining the nutrient levels in litter, but climate and the composition of the mineral soil, parent material and the humus are also of importance. Thus, in undisturbed Scots pine systems it appears that the concentration of N may be related to both climate (Berg et al. 1995a) and N concentration in the humus (Berg et al. 1999b).

This chapter focuses on the litter fall from trees and aims to give an insight into the present knowledge on the chemical composition of needle litter fall in mainly pine stands, though other conifers and deciduous species are included. Though the emphasis is on foliar litter, we will include some discussion on wood. We introduce the characteristics of the main insoluble chemical constituents of plant litter. We also identify factors that may influence litter chemical composition, with an emphasis on climate and soils. As data are limited, case studies have been used. The purpose of this chapter is not to explain the chemical composition of litter as based on the physiology of living plants, but rather to take into account the effect of general environmental conditions. As such, we focus on environmental components that appear to have a significant influence on the rate and pattern of litter decay and humus formation.

4.2 Organic-chemical components of plant litter and fiber structure

4.2.1 Organic-chemical components

Together with lignin, which is a complex polymer compound formed by aromatic rings (Fig. 3.3), the polymer carbohydrates form the plant fiber structures. The quantitatively most common components in plant litter are the different polymer carbohydrates such as cellulose and the main hemicelluloses. Of these, cellulose, the most common component, is made up of glucose units connected with β-1-4 bonds forming long chains of molecules organized into fibers. Cellulose may constitute between 10 and 50% of the litter mass (Table 4.1).

Hemicelluloses are polymers of sugars other than glucose that also form long chains of molecules that are built into the fiber. Hemicelluloses are named based on the simple sugars they are synthesized from, such as mannan, galactan, arabinan and xylan. The proportions of hemicelluloses vary among litter species (Table 4.1). Differences in the major hemicelluloses are primarily reflected in the concentrations of xylose and mannose (Eriksson et al. 1990). Deciduous leaves are lower in mannans whereas Norway spruce needles have higher levels, and birch leaves are richer in xylans. The ratio of hemicelluloses to cellulose ranges from about 0.7 to 1.2 (Table 4.1) with higher ratios often seen in deciduous litter as compared to coniferous litter. Hemicelluloses may together make up as much as 30-40% of the fiber and are normally present at between 1 and 10% each (Table 4.1). In contrast with cellulose, hemicelluloses are often branched.

Lignin often makes up between 15 and 40% of the litter mass. In some extreme cases, litter can have lignin contents as low as 4% or as high as 50%. Lignin, in contrast with cellulose, is a highly variable molecule. The initial composition of lignin varies with the plant species, and the variation is enough to make the lignin of each species unique. This also rules the terminology. Thus, the native lignin of different species may be specified by the name of the species, for example, Norway spruce lignin and aspen lignin. A generalized structure of Norway spruce lignin is illustrated in Fig. 3.3. The terminology pertaining to lignin and its transformation products is, however, not always clear; especially after some degradation has taken place (Dean 1997, see Chap. 2 and Glossary).

The types of lignin formed in coniferous and deciduous trees are different. Whereas deciduous species contain varying ratios of syringyl and guaiacyl types of lignin, conifers have mainly guaiacyl lignin (Fengel and Wegener 1983). While some basic structural elements are common over a wide range of species, individual species show variation among a variety of groups such as methoxyl groups and other substituents located at different sites in the molecule.

The lignin content of angiosperms (hardwood) is generally lower than that of the gymnosperms (softwood), although the range varies in both groups. The lignins of hardwoods contain more sinapyl alcohol units, each having two methoxyl groups. If a methoxyl group occupies the para position on the benzyl alcohol subunit instead of a hydroxyl group, the molecule is significantly more susceptible to enzymatic oxidation (Ander and Marzullo 1997).

Table 4.1. Comparison of the major organic-chemical compounds in several boreal litter types. Foliar litter data from Berg and Ekbohm (1991), wood data from Eriksson et al. (1990)

Litter type	Concentration of compound [mg g^{-1}]									H:C
	Wsol	Esol	Lig	Glu	Man	Xyl	Gal	Ara	Rha	
Coniferous needles										
S. pine (br)	164	113	231	245	75	23	32	36	3	0.69
S. pine (gr)	180	90	220	217	75	19	30	23	2	-
LP. pine (br)	103	42	381	254	90	34	46	48	6	0.88
LP. pine (gr)	147	36	346	239	81	29	37	40	33	-
N. spruce (br)	32	48	318	288	105	33	28	40	7	0.74
N. spruce (gr)	-	-	-	-	-	-	-	-	-	-
Deciduous leaf litter										
S. birch (br)	241	57	330	166	14	77	44	49	16	1.2
S. birch (gr)	222	57	322	168	9	54	51	36	12	-
G. alder	254	39	264	116	10	30	32	44	9	1.08
Deciduous wood										
S. birch	-	-	217	351	9	207	-	-	-	-
Aspen	-	-	220	462	16	189	-	-	-	-
R. alder	-	-	246	470	4	176	-	-	-	-
Coniferous wood										
S. pine	-	300	383	111	65	-	-	-	-	-
R. pine	-	279	449	123	84	-	-	-	-	-
N. spruce	-	271	416	136	52	-	-	-	-	-

Wsol Water soluble, *Esol* Ethanol soluble, *Lig* Lignin, *Glu* Glucans (cellulose), *Man* Mannans, *Xyl* Xylans, *Gal* Galactans, *Ara* Arabinans, *Rha* Rhamnans, *S. pine* Scots pine, *LP pine* lodgepole pine, *R. pine* red pine, *N. spruce* Norway spruce, *S. birch* silver birch, *G. alder* grey alder, *R. alder* red alder. *H:C* Hemicellulose: Cellulose ratio, *(br)* brown litter, *(gr)* green litter.

Litter contains large quantities of low-molecular weight substances too, such as amino acids, simple sugars, lower fatty acids and lower molecular weight phenolic substances. Complex compounds such as high-molecular weight fatty acids and complex phenolic compounds are also found. We may be able to identify some hundred different molecules from these two groups. Often they are analyzed as water-solubles for the former group and ethanol- or acetone-solubles for the latter.

Cutin and suberin are resistant molecules that can influence decomposition and increase in concentration during decay (Kolattukudy 1980, 1981). Although present in rather small amounts and seldom identified in litter decomposition studies, these polyesters act as barriers to protect living plants and to delay invasion by microorganisms. Cutin is found on and in leaves and suberin in bark and roots. Both are polymers composed of hydroxy- and epoxy-alkonic acids. Kögel-Knabner et al. (1989) extracted these acids from the L layers of European beech and Norway spruce forests. They found that cutin and suberin contributed 12 to 24 mg g^{-1} of the organic matter.

4.2.2 Fiber structure

The insoluble components of plant litter are concentrated in the cell wall, a multi-layered structure. The wood cell wall is composed of various layers (Fig. 4.1) and is made up of a primary wall (P) and the secondary wall (S), which has three layers designated S1, S2, and S3. The middle lamella and the primary wall make up the compound middle lamella that is located between the secondary wall layers of adjacent cells (Core et al. 1979). The S3 layer is located closest to the lumen (L). The normally thickest layer (S2) is the middle layer and S1 is the outermost layer of the secondary wall. These layers are distinct from each other because the cellulose is arranged in different microfibrillar orientations. A model describing the arrangement of lignin, hemicellulose, and cellulose within the cell wall was proposed (Kerr and Goring 1975, Fig. 4.1). The model shows how a matrix of lignin and hemicellulose encrusts the cellulose fibrils. There is a tremendous diversity in wood structure among the hundreds of hardwood species that grow throughout the world (Panshin and de Zeeuw 1980). Still, the drawing in Fig. 4.1 may serve as a model for our discussion.

In plant fibers the cellulose, the hemicelluloses and the lignin molecules are not only combined physically, but the celluloses are normally more or less encrusted with lignin. Within the cell wall, cellulose forms microfibrils composed of individual strands of cellulose that are often about 10-25 nm in diameter (Fig. 4.1). Microfibrils group together into larger strands called macrofibrils. These are visible with a light microscope and are about 0.5 μm in thickness.

Of the several cell wall layers the thickest (S2) normally has a width of 0.5 - 4.0 μm. In their turn, the walls are constructed of a matrix of cellulose, hemicellulose, and in many plant tissues, lignin. The thickness of the entire primary and secondary wall complex is highly variable.

The formation of lignin in the fibers (lignification) of the live plant is a slower process than the formation of cellulose. As a result, the last parts of the cell wall to be formed may be very low in lignin and the older parts richer. Thus, in wood, lignin is distributed throughout the secondary (S) wall and compound middle lamella, but the greatest concentration is in the middle lamella. The secondary wall makes up a large part of the total cell wall and most of the cell wall lignin (60-80%) is located in this region (Musha and Goring 1975; Saka and Thomas 1982a,b). The distribution of hemicellulose parallels that of lignin within the wall (Parameswaran and Liese 1982). The hemicellulose surrounds the cellulose microfibrils and occupies spaces between the fibrils.

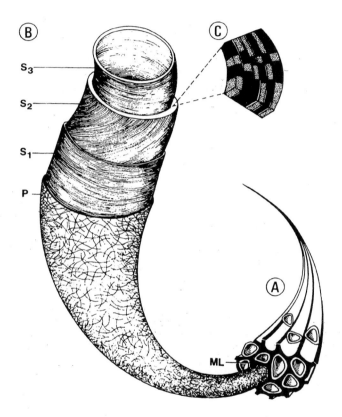

Fig. 4.1. Overview of a plant fiber. **A** Tracheids. **B** Cell wall layers. **C** Arrangements of polymer carbohydrates and lignin in the secondary wall. Middle lamella (*ML*), primary wall (*P*), layers of the secondary wall (*S1, S2, and S3*). The microfibrillar orientation and the thickness are different among the layers and some species have an additional warty layer over the inner (*S3*) layer. The model demonstrates the distribution of the lignin-hemicellulose matrix (*black*) hemicellulose (*white*) and cellulose fibrils (*dotted*). (From Eriksson et al. 1990)

4.3 Nutrient concentrations in newly shed litter

4.3.1 General features

The nutrients found in newly shed plant litter have their origins in strictly controlled structures in the live plant parts, and a nutrient like N can be found in membranes, cytoplasmic enzymes, structural proteins or nucleic acids. In green foliage, ribulose bisphosphate carboxylase/oxygenase (rubisco) may account for the majority of the N. When a leaf dies and foliar litter is being formed, these cellular constituents disintegrate, at least in part. Before a leaf dies, a proportion of the nutrients is translocated into the perennial portion of the plant, leaving the remainder in the dead material, a process that may take place over several weeks.

Nutrients, as usually measured by ecologists, are expressed in their elemental or ionic form, making no distinction as to the origin of the nutrient within the plant's structure. What is often measured as just a "mineral nutrient," for example N, can thus originate from a number of different components in the litter, such as proteins or nucleic acids. Nitrogen is found in concentrations between 0.2 and 3.0% in foliar litters. In woody structures, such as branches, the concentration may be as low as 0.02%. Phosphorus is bound in nucleic acids and S is found in proteins too, among other molecules. We cannot assume that e.g. the total N in different litter species is chemically bound in the same molecules across species and concentration ranges.

When decomposition and microbial ingrowth have started, further changes occur and the distribution of nutrients in compounds, as well as concentrations of nutrients, will be very different from both the living and the freshly senescent material. In this book we will not generally discuss the nutrients in terms of the macromolecules they are part of, but rather simply as nutrients.

4.3.2 Nutrient resorption

The chemical composition of the living plant is reflected also in its litter. This applies to several components such as lignin, the relative composition of hemicelluloses and, to a smaller extent, nutrients.

Many genera, such as pine, growing on relatively nutrient-poor soils, retrieve the main part of their nutrients before shedding their foliar litter. This "inner circulation" is a conserving mechanism that has been suggested to be in effect mainly on nutrient-poor soils (Gosz 1981; Staaf and Berg 1981). An example at the opposite extreme is the N_2-fixing genus alder (*Alnus*), which produces leaf litter that has as high a concentration of N as the live leaves, ca. 2-3%.

Trees withdraw substances other than nutrients before shedding their leaves and needles. Thus, at the same time, different soluble C components, such as sugars and phenolics, are withdrawn resulting in a mass loss from the living tissue. This may result in an increase in concentration of those nutrients that are withdrawn to a lesser extent, and a decrease for those that have been withdrawn to a greater extent. Thus, nutrients that have an increase in concentration during senescence may not represent a true increase in amount but rather an increase in proportion as total organic compounds are depleted.

A thorough study on leaves of European beech (Staaf 1982), indicates that there was a positive correlation between the withdrawal of nutrients and the concentration of the nutrient in green leaves. This relationship was especially steep for N (Fig. 4.2), showing that a higher initial concentration led to a large increase in withdrawal. In contrast, the relationship was rather flat for Ca, indicating a lower effect of initial concentration on the withdrawal (Fig. 4.2). Soil pH had a negative effect on the withdrawal of Ca (Staaf 1982) and at sites with a lower soil pH there was a lower withdrawal. This effect was seen only for Ca.

A comparison was made between concentrations of the main nutrients in green leaves collected in July as compared to newly shed ones (Table 4.2). It was found that concentrations of N, P, S and K were considerably lower in newly shed litter, whereas Ca, Mg and Mn had increased concentrations in the shed litter as compared to the green leaves. For the heavy metals, the concentrations were the same or higher than the initial values, based on a limited amount of data.

In the case of Scots pine, lodgepole pine, silver birch, and trembling aspen the concentrations of N may decrease to about 1/3 of that in green leaves before the leaves and needles are shed in the autumn. For example, in pines the concentration may decrease from about 12-14 mg g^{-1} to about 3-4 mg g^{-1} (Table 4.2). This retrieval process may of course be disturbed, occasionally leading to extreme levels of N (Table 4.3). For Norway spruce, the N concentration decreased to about 50% of the level in green needles.

When just comparing concentrations of remaining P to those in green leaves they were of the same magnitude as for N, from 15 to 50%. For S, less of the nutrient was retrieved, and in general more S remains with concentrations ranging from 38 to 103% of those in green leaves. For K there was a difference between coniferous and trees, with deciduous trees having clearly higher concentrations when shed, in the range of 40-50% of the concentrations in green foliage. The newly shed conifer needles had less than 25% of their summer K concentrations, while pines as a group had even lower levels.

Calcium was not retrieved very much, which results in an increase in concentration to 115-220% (see below). The remaining Mg ranged from 43 to 113% of the initial concentration. Manganese contrasted to the other nutrients by having increased concentrations over all cases ranging from 123 to 250%.

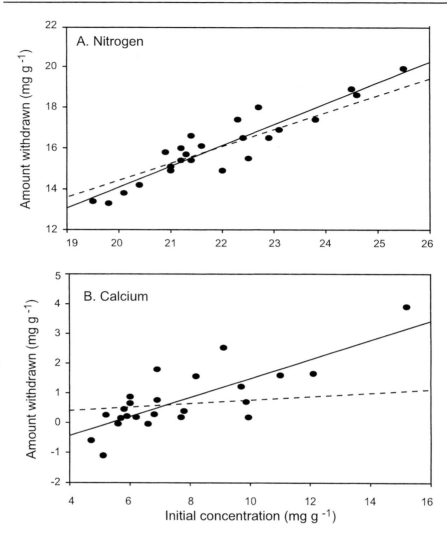

Fig. 4.2. Linear relationships between concentrations of N and Ca in green leaves of European beech and the amount of the nutrient that has been withdrawn before senescence. The amount withdrawn is expressed per leaf weight before senescence. The full line shows that the amount withdrawn increases with the initial concentration in the green leaves. The dashed line shows a theoretical relationship in case a constant fraction should have been withdrawn. (Redrawn from Staaf 1982)

Table 4.2. Comparison of concentrations of some common nutrients in green leaves collected in July and in the corresponding foliar litter collected at litter fall (B. Berg unpubl.). N.B. The table compares only concentrations and does not consider retention or withdrawal of nutrients (Sec. 4.3.2). Data for European beech from Staaf (1982)

Litter type	Concentration of nutrient [mg g^{-1}]						
	N	P	S	K	Ca	Mg	Mn
S. pine (br)	3.6	0.20	0.44	0.5	5.6	0.34	1.19
S. pine (gr)	12.1	1.36	0.81	5.9	3.9	0.79	0.53
% remaining	30	15	55	8	143	43	224
LP. pine (br)	3.1	0.29	0.44	0.5	8.7	1.06	2.03
LP. pine (gr)	10.5	0.82	1.17	3.8	4.0	0.93	0.82
% remaining	30	35	38	13	220	113	250
N. spruce (br)	4.2	0.41	-	1.0	13.1	0.89	1.32
N. spruce (gr)	8.5	1.32	-	4.0	11.3	1.22	1.07
% remaining	49.0	31	-	24	115	73	123
S. birch (br)	7.7	1.05	0.80	4.7	11.8	3.30	1.23
S. birch (gr)	24.3	1.96	1.54	9.0	9.5	3.37	0.76
% remaining	32	53	52	52	124	98	158
T. aspen (br)	6.8	0.63	1.37	6.3	17.1	2.13	0.15
T. aspen (gr)	24.2	2.12	1.87	14.2	8.4	2.29	0.10
% remaining	28	30	73	44	204	92	150
E. beech (br)	9.1	0.63	1.21	2.7	10.0	1.70	-
E. beech (gr)	22.6	1.44	1.18	5.4	7.7	1.67	-
% remaining	40	44	103	50	130	102	-

S. pine Scots pine, *LP pine* lodgepole pine, *N. spruce* Norway spruce, *S. birch* silver birch, *T. aspen* trembling aspen, *E. beech* European beech, *br* brown, *gr* green.

4.3.3 Nutrients in Scots pine needle litter - a case study

Extensive data are available on the initial organic chemical and nutrient composition of Scots pine needle litter over extended periods of time and a wide geographical area. Using this data, we can examine the variability of chemical composition within a single stand over time, among stands in a small geographical area and among stands across a climatic gradient.

Annual variation at one site

In a single stand there is a clear variation in the chemical composition of the newly shed needle litter over different years. This is illustrated (Table 4.3) with an investigation in which some nutrients in freshly fallen needle litter were measured in 17 consecutive years. Also, the levels of water solubles and lignin were followed, with some constituents being much more variable than others.

Table 4.3. Annual variation in concentration of solubles, lignin and nutrients of Scots pine (*P. silvestris*) needle litter collected in a nutrient-poor forest at Jädraås (SWECON site, cf. Appendix III) (Johansson et al. 1995)

Year	Concentration [mg·g⁻¹]									
	Wsol	Esol	Lig	N	P	S	Ca	K	Mg	Mn
1973	92	120	223	3.8	0.19	0.42	6.5	0.73	0.38	1.55
1974	145	84	276	4.2	0.22	0.29	5.4	0.71	0.49	n.d.
1975	172	107	238	3.4	0.20	0.32	4.7	0.61	0.39	n.d.
1976	151	89	255	4.0	0.21	0.36	4.9	0.53	0.42	n.d.
1977	202	102	224	4.1	0.19	0.38	6.0	0.87	0.42	1.02
1978	164	96	257	3.8	0.21	0.33	5.5	0.62	0.55	1.00
1979	129	95	288	10.4	0.29	0.78	2.3	0.97	0.39	0.31
1980	180	102	246	3.8	0.18	0.50	6.1	1.72	0.53	0.77
1981	213	94	231	3.9	0.28	0.61	7.1	1.02	0.58	1.17
1982	164	113	231	4.8	0.33	0.55	4.4	1.07	0.49	0.79
1983	178	112	229	3.8	0.30	0.45	5.9	0.9	0.39	1.08
1984	82	116	288	3.7	0.21	0.47	6.3	0.82	0.44	1.12
1985	182	94	241	3.0	0.19	0.45	4.8	0.52	0.38	1.24
1986	170	89	257	4.0	0.23	0.44	5.6	0.58	0.57	1.13
1987	162	100	250	3.8	0.21	0.42	4.9	0.55	0.41	1.18
1988	165	94	247	3.8	0.21	0.39	5.0	0.67	0.38	1.18
1989	n.d.	n.d.	n.d.	3.6	0.17	0.38	4.0	0.59	0.42	0.92
AVG	159	100	249	4.2	0.23	0.44	5.3	0.79	0.45	1.03
S.D.	35	11	21	1.6	0.05	0.12	1.1	0.29	0.07	0.27

Wsol Water solubles, *Esol* ethanol solubles, *Lig* lignin, *n.d.* not determined, *AVG* average value, *S.D.* standard deviation

Concentrations of N varied from 3.0 mg g⁻¹ up to a very high value of 10.4 mg g⁻¹. Compared with other years, the latter value is exceptionally high in relation to concentrations of elements such as P and S in the same year. The frequency of occurrence of such a high value has not been established and may be regarded as a consequence of an unknown extreme event. However, in another study on litter chemical composition over a transect, a similarly high value was found (Johansson et al. 1995). With the next highest value being 4.8 mg g⁻¹ (Table 4.3), N concentration still shows considerable variation (1.6 fold) when compared to the lowest value. The average concentration for N at this site was 4.2 mg g⁻¹. The concentration of P varied between 0.17 and 0.33 mg g⁻¹ showing a range factor of 1.9. For S, the range was from 0.29 to 0.78 mg g⁻¹, giving a factor of 2.7.

Within the same stand, Ca concentrations varied with a factor of 3.1 from 2.3 up to 7.1 mg g⁻¹, a relatively large variation considering that earlier analyses indicate a strong dependence of soil type for this nutrient. For K, the mean value was 0.79 mg g⁻¹ and the range from 0.52 mg g⁻¹ to 1.72 mg g⁻¹, thus with a factor of 3.3. The mean for Mg was 0.45 mg g⁻¹, and the range was from 0.38 to 0.58 mg g⁻¹, a range with a factor of 1.5. Manganese was the most variable nutrient; the mean value was 1.03 mg g⁻¹ and the range from 0.31 to 1.55 mg g⁻¹, thus a range factor of 5.

The concentrations of the main nutrients N, P and S were in the average proportions of 1:0.055:0.105. As we will discuss later, N and P have been ascribed the role of being rate limiting for decomposition. When we relate both P and S to N the relative proportions of P are seen to vary considerably, from 0.028 to 0.079, and for S from 0.069 to 0.156. There was a variation in proportions between years that may influence which nutrient was rate limiting (see Chaps. 2 and 6). Even disregarding the few extreme values, there is still clear variation between years. A trend analysis did not reveal any significant change in nutrient concentrations over time. The variation in concentrations of water-soluble substances ranged from 82 to 213 mg g^{-1}, with an average value of 159 mg g^{-1}. For lignin concentrations, the range was from 223 to 288 mg g^{-1} with an average value of 249 mg g^{-1}.

No strong correlation existed among the constituents and, using the Spearman rank correlation, only three correlations were significant: water solubles and lignin ($r = -0.535$, $p = 0.033$), N and P ($r = 0.546$, $p = 0.029$), and S and K ($r = 0.599$, $p = 0.014$). Ash concentrations in the collections of Scots pine needle litter were relatively low (average value = 20 mg g^{-1}) as compared with those of other tree species (Bogatyrev et al. 1983).

Variation among Scots pine sites and in a transect

In Europe, Scots pine grows mainly from the Barents Sea in the north to the Pyrenees and northern Greece in the south (Fig. 4.3), although it forms forests to about the latitude of the Alps and the Carpathians. Scots pine may grow on both nutrient-poor granite sand and on clayey soils. On a European scale, the magnitude and pattern of litter fall varies with the geographical position and climate.

Litter fall begins relatively early in the north, close to Barents Sea (ca. 70° N) normally in the first week of August. In south Poland (ca. 50° N), it may start as late as November. A strong drought may change this pattern and induce an earlier litter fall. That drought does influence the onset of litter fall is seen in the Mediterranean climate where native pine species shed their needles in July. Scots pine, when growing in a Mediterranean climate, has adopted the litter-fall pattern of the Mediterranean pine species, and has its main foliar litter fall in summer during the dry period.

The litter chemical composition varies with the site's geographical position and climate (Berg et al. 1995a). A study along a transect ranging from Barents Sea in the north to about the Carpathian Mountains in the south, thus encompassing half the length of Europe (Fig. 4.3), shows a clear trend in chemical composition with climate. Concentrations of N, P, S and K are positively related to AET, for example, N levels range from about 3 mg g^{-1} in the north to about 9 mg g^{-1} in the more southern locations (Fig. 4.4).

For Mn, a significant but negative relationship indicates that the highest Mn concentrations were found in the north at lower AET, and the lowest at the southern sites with a higher AET.

This gradient in nutrient levels appears to be normal in pines as a genus (cf Berg and Meentemeyer 2002). The progressive increase in concentrations of N, P, S and K in litter with increasing AET and decreasing latitude may very well

Fig. 4.3. Map giving the main distribution of Scots pine in Europe. Note that Scots pine occurs in plantations and as single plants, for example, in the Apennines and the Pyrenees. (Redrawn from Anonymous 2002)

reflect a direct or an indirect effect of climate on litter quality. This may indicate that a "better" climate gives more nutrient rich green leaves and needles. With a fraction of the nutrients being retained in the litter the result is higher concentrations in the shed litter.

Within Scots pine, the variation in litter N concentration is weakly but positively correlated to the concentration of N in the organic matter in the humus layers (R^2= 0.121; n=36). This ties to climate and illustrates the triangular relationship between soil, climate and litter discussed by Aerts (1997). Thus, if a climate favorable for the growing plant gives a higher N concentration in the needles and in the needle litter, this leads to humus with a higher N concentration (see Chap.11) and thus a nutrient feedback to the plants. As this relationship was seen for N, we may expect that it is valid also for P and S.

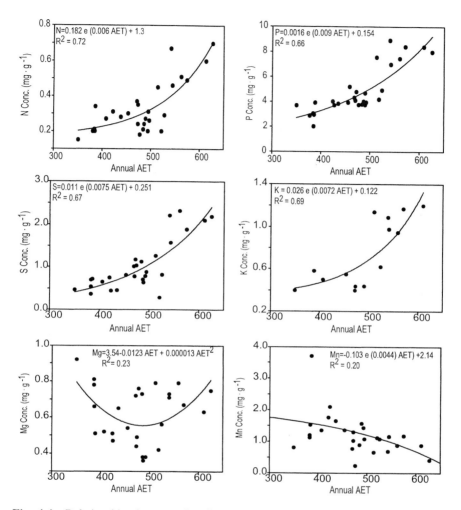

Fig. 4.4. Relationships between the climatic index actual evapotranspiration (AET) and concentrations of nutrients in newly shed Scots pine needle litter. N, P, S, K, Mg, and Mn. The litter was collected in a climatic transect of Scots pine forests ranging from Barents Sea in the north to central Europe. (Berg et al. 1995a)

4.3.4 Several deciduous and coniferous leaf litters

To expand this review of litter chemical composition from Scots pine to even the most common litter types in a given ecosystem is impossible today, due to a lack of basic information. We can, however, distinguish some of the main patterns that relate to tree genera or species and their nutritional patterns.

Litter chemical composition may be related to a few factors such as soil nutrient supply, climate (see Fig. 4.4) and species. A set of data including average values from some principal northern European tree species will indicate clear patterns in nutrient richness among genera/species. There are general differences between deciduous and coniferous trees (Table 4.4), one difference being the contents of hemicelluloses and lignin (Table 4.1) and there are also clear differences among species. The coniferous foliar litter overall appears less rich in nutrients such as N, P, Ca and K than the deciduous ones (Table 4.4). The eight coniferous litter species in general had N levels under 7 mg g^{-1} whereas the deciduous ones were higher. Likewise, there were larger differences in P levels, with deciduous litters averaging three times as much P as coniferous litters. For Ca, the pines were generally low, Norway spruce had a higher value, and deciduous trees had Ca concentrations that were at least 2-10 times higher than those of coniferous litters, with the average value about 2.5 times higher. Mg and Mn were more similar among species, but there was a tendency for concentrations of Mn to be lower in the deciduous leaves.

Table 4.4. Major nutrients in selected boreal and temperate litter species

Litter	Concentration of nutrient [mg g^{-1}]						
	N	P	S	K	Ca	Mg	Mn
Coniferous needles							
Scots pine[a]	4.8	0.33	0.55	1.07	4.4	0.49	0.79
Lodgepole pine[b]	3.9	0.34	0.62	0.56	6.4	0.95	1.79
Maritime pine[c]	6.8	0.54	1.01	1.95	3.1	1.90	0.59
Red pine[c]	6.0	0.36	0.73	1.40	8.9	2.00	0.73
White pine[c]	5.9	0.21	0.68	0.70	7.2	1.10	0.80
Jack pine[c]	7.8	0.64	0.77	2.30	4.0	2.10	0.25
Limber pine[c]	4.3	0.43	0.52	1.10	5.3	1.10	0.21
Norway spruce[d]	4.9	0.45	0.73	0.72	17.9	0.65	-
AVERAGE	5.6	0.41	0.70	1.23	7.2	1.29	0.74
Deciduous leaves							
Grey alder[e]	30.7	1.37	6.12	15.6	12.3	2.32	0.10
Silver birch[e]	7.7	1.05	0.80	4.7	11.8	3.30	1.23
Ash[f]	8.6	1.96	-	15.3	33.2	2.28	0.03
Mountain ash[f]	7.1	0.31	-	10.8	12.4	2.86	0.30
Trembling aspen[f]	8.2	0.93	-	5.1	29.9	4.69	0.53
European maple[f]	5.1	3.15	-	13.1	20.4	1.46	0.12
European beech[d]	9.5	1.40	1.30	2.3	7.4	1.20	1.90
AVERAGE	10.9	1.45	2.74	9.6	18.2	2.59	0.60

[a]Johansson et al. (1995).
[b]Berg and Lundmark (1987).
[c]C. McClaugherty and B. Berg (unpubl.).
[d]Berg et al. (2000).
[e]Berg and Ekbohm (1991).
[f]Bogatyrev et al. (1983).

Table 4.5. Average values for some chemical components in some needle litter types collected in paired stands in boreal and temperate forests. Significant differences are indicated. (Reurslag and Berg 1993)

	Wsol	Esol	Lig	N	P	Ca	Mg	K	Mn
Components [mg g^{-1}]									
Scots pine - lodgepole pine, data from three paired stands									
S. pine									
Avg	148	99.4	256.8	3.9	0.27	5.8	0.56	0.67	1.56
SD	44	13.2	25.8	0.5	0.06	0.9	0.10	0.36	0.38
n	15	15	15	7	7	7	7	7	7
LP. Pine									
Avg.	125	47.9	360	3.9	0.37	6.9	0.97	0.57	2.40
SD	34	6.6	19	0.4	0.06	1.1	0.078	0.16	0.399
n	16	16	16	16	7	7	7	7	7
Sign. (*p*)	ns	0.001	0.001	ns	0.01	ns	0.001	ns	0.01
Scots pine - Norway spruce, data from eight paired stands									
S. pine									
Avg.	132	97.4	261	4.1	0.25	6.0	0.53	0.81	1.20
SD	43	12.0	40	0.2	0.06	1.6	0.16	0.25	0.48
n	8	8	8	8	8	8	8	8	8
N. spruce									
Avg.	153	48.9	305	5.2	0.75	15.2	1.06	1.90	2.45
SD	36	8.2	23	0.4	0.27	9.2	0.28	0.65	1.02
n	8	8	8	8	8	8	8	8	8
Sign (*p*)	0.001	ns	0.05	0.01	ns	ns	0.001	0.01	ns
Scots pine - silver birch, data from one pair of stands									
S. pine									
Avg.	159	100.0	249	4.2	0.23	5.3	0.45	0.79	1.03
SD	35	11.0	21	1.6	0.05	1.1	0.07	0.29	0.28
n	17	17	17	17	17	17	17	17	17
S. birch									
Avg.	275	48.7	294	8.1	1.61	12.6	3.28	5.59	2.40
SD	34	11.1	28	0.7	0.57	1.4	0.55	1.58	0.88
n	3	3	3	3	3	3	3	3	3
Sign (*p*)	0.001	0.001	0.05	0.001	0.001	0.001	0.001	0.001	0.05
Norway spruce - silver birch, data from paired stands									
N. spruce									
Avg.	153	48.9	305	5.2	0.75	15.2	1.06	1.90	2.45
SD	38	8.2	23	0.4	0.27	9.2	0.65	0.65	1.02
n	8	8	8	8	8	8	8	8	8
S. birch									
Avg.	231	66.4	284	10.2	0.61	15.3	3.29	5.26	2.47
SD	10	14.5	7	2.4	0.17	3.5	0.18	1.56	1.49
n	3	3	3	3	3	3	3	3	3
Sign (*p*)	0.001	ns	0.05	0.01	ns	ns	0.001	0.01	ns

Wsol Water solubles, *Esol* ethanol solubles, *Lig* lignin, *Avg* average value, *SD* standard deviation, *Sign (p)* significant differences tested by t-test, *n* number of samplings, *ns* not significant

Generally, the most nutrient-poor foliar litters were those of pines, for example Scots pine, lodgepole pine, red pine and white pine. This general picture applies to most of the measured nutrients (N, Mg, K, Mn). For N, ash, aspen and beech had similar levels, and for P, mountain ash and the conifers had similar levels.

A set of Swedish data originates from stands growing on soils and at climates representative for the species, and the ranking of species is confirmed by paired comparisons made on the same soil and climate, where needles of Scots pine were generally more nutrient poor than those of Norway spruce. Spruce needles tend to be less nutrient rich than birch leaf litter (Table 4.5).

A study on concentrations of K in needle litter (Laskowski et al. 1995) encompassed more than 25 boreal and temperate tree species. The full range for K concentrations in the whole data set was 0.31-15.64 mg g^{-1}. They found a large and statistically significant ($p<0.0001$) difference in average initial K concentrations between coniferous and deciduous litters (1.03 vs. 4.52 mg g^{-1} respectively). Of individual species, lodgepole pine needle litter had the lowest initial concentrations among the litter types studied, followed by those of Scots pine. Both these litter types had lower initial K concentrations than were found in the leaf litter of Norway spruce, oak-hornbeam and birch. The highest average K value was that for grey alder leaves (8.26 mg g^{-1}) followed by that for silver birch leaves (5.01 mg g^{-1}). In contrast, leaves of European beech, with 1.67 mg g^{-1}, were in the same range as the coniferous litter. Their investigation also covered temperate forests and covered both the most common litter species found in forests of north-central Europe, as well as some major North American species.

Influence of soils

Soil chemical composition, including the availability of nutrients, has an influence on the chemical composition of leaves and thus also on chemical composition of foliar litter. The chemical composition of the soil and the availability of the nutrients would be two main factors acting either directly or indirectly on the composi-

Table 4.6. Concentrations of nutrients in leaf litter of European beech on two different types of forest floor. The litters were sampled from stands within a limited region of about 100 x 100 km. (Staaf 1982)

	N	P	S	Ca	Mg	K	N:P:S
	\multicolumn{7}{c}{Concentration [mg g^{-1}]}						
			\multicolumn{4}{c}{Mor humus n=14}			
Mean	9.06	0.66	1.18	8.54	1.50	2.59	1:0.07:0.13
SD	0.43	0.10	0.02	0.85	0.07	0.49	
			\multicolumn{4}{c}{Mull humus n=10}			
Mean	9.04	0.58	1.26	11.97	1.97	2.84	1:0.06:0.14
SD	0.33	0.04	0.05	6.3	0.04	0.31	
			\multicolumn{4}{c}{t-test}			
t-Value	0.07	0.50	-1.12	-4.73	-4.82	-0.93	
Significance (p)	ns	ns	ns	<0.001	<0.0001	ns	

n number of sites sampled, *ns* not significant

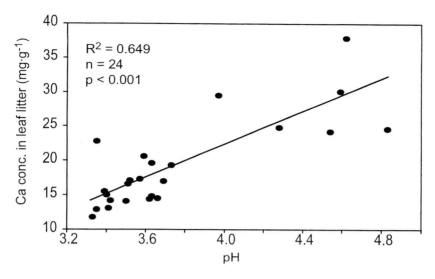

Fig. 4.5. A relationship between pH in the humus (A_0 to A_{01}-A_{02}) layers and Ca concentration in the shed leaf litter of European beech. (Staaf 1982)

tion of foliar litter. That nutrient availability has an influence is seen from, for example, N fertilizer experiments (Sect. 4.5.1) in which different dosages of ammonium nitrate resulted in increases in the N concentration in the needle litter. Nitrogen availability in undisturbed or unmanaged forests may vary greatly across sites dominated by different species (Pastor et al. 1984). However, variations in N availability among natural, undisturbed sites, which are dominated by the same species, appear to be much smaller. For a group of species the variation in, for example, litter N concentration can be related to the concentration of N in the organic matter in the humus layers ($R^2 = 0.668$; $n=61$)(Berg et al. 1999b). In a study on litter nutrient levels among 24 plots within an area of less than 100×100 km Staaf (1982) showed limited differences in Ca and Mg levels among European beech stands (Table 4.6). In another study, Johansson et al. (1995) compared N levels of needle litter among three nearby stands of Scots pine with the same climate but with different soils; on a dry sandy sediment soil; on a richer till soil; on a moist mull soil on clay. The N concentrations increased in order 2.9, 3.0, and 3.2 mg g^{-1}, respectively. Thus, in this case, soil type had only a minor effect on needle litter N content, as compared to that of climate.

Although the levels of N available to plants in natural soils often are low enough not to influence the litter chemical composition, the levels of other nutrients in litter appear to be more directly dependent on their occurrence in soil. Thus, the concentrations of Ca and Mg in litter appear to increase as their availability in soil increases. These and other nutrients such as Mn are dependent on pH for their mobility. At lower pH values a better supply of mobile Mn will lead to higher levels in the leaves and needles.

The influence of soil pH is also well illustrated by a study on beech leaves in 24 stands in a climatically homogeneous area. Ten plots with mull soils had significantly higher humus pH and the litter had higher concentrations of Ca (12.0 vs. 8.5 mg g^{-1}, t=4.722, $p<0.002$) and Mg (1.97 vs. 1.56 mg g^{-1}, t=4.818, $p<0.0001$) than those with a mor soil, whereas the concentration of N, P, S and K were not affected. In the same study on Ca in beech leaves, Staaf (1982) found a clear relationship between the humus (A_0) pH and Ca concentrations in the leaf litter (Fig. 4.5).

Another group of nutrients has an indirect effect on the chemical composition of litter. For example, a lack of B will not only be reflected in a lower B concentration in the litter, but will also have an indirect effect by influencing the litter lignin level. Boron has an important role for the formation of an enzyme transporting phenols out from the needles. A lack of B results in an accumulation of phenolics in the needles, which causes increased lignin synthesis (Lewis 1980; Dugger 1983, see also Chap.12). Excessively high levels of Cu have been suggested to have a similar effect.

Variation over a climatic transect

Species other than Scots pine also appear to have a variation in chemical composition with climate. Available data for different pine species follows the same pattern as for Scots pine (Fig. 4.6a). Furthermore, using all available data for foliar litter, similar patterns of N concentrations across a gradient of AET were shown (Berg and Meentemeyer 2002). Thus, all available data for N in foliar litter on a European basis have been related to AET, indicating a more general relationship (Fig. 4.6). In that comparison, N_2-fixing species were excluded. Considering this wider range of species, the relationships are weaker than with Scots pine alone, but the trends remain the same for both N and S. For Mn there was a similar general negative relationship to AET as there was for Scots pine and no general relationship was found for other nutrients including P, Ca, Mg and K.

4.3.5 A comparison of some common species

There are limited comparative studies on the chemical composition of some common boreal foliar litter types. A comparison of Scots pine and Norway spruce needle litter over eight paired stands (Table 4.5) showed significant differences between them, except for the labile fraction of water-soluble substances. Norway spruce needle litter was richer in all the measured nutrients (N, P, Ca, Mg, K and Mn) as well as in lignin. On the other hand, the ethanol-soluble fraction was larger in the litter of Scots pine.

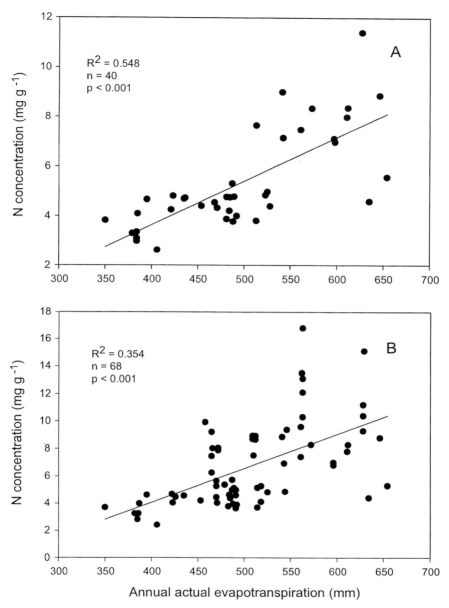

Fig. 4.6. The relationship between climate, as indexed by actual evapotranspiration (AET), and N concentration in foliar litter. **A** Different pine species at sites across Europe. **B** All available data on coniferous and deciduous litter across Europe. (Berg and Meentemeyer 2002)

In another comparison between paired stands of Scots pine and lodgepole pine, the differences had a different pattern (Table 4.5). Lodgepole pine had higher levels of P, Mg, Mn and lignin, whereas Scots pine was richer in ethanol-soluble substances. Nitrogen, K and Ca were not significantly different.

A similar comparison made between silver birch and Scots pine litters collected at Jädraås (site described in Appendix III), showed that silver birch leaf litter was significantly richer in all the analyzed nutrients (Table 4.5). Concentrations of water solubles and lignin were also higher in the birch leaf litter. On the other hand, the fraction with ethanol solubles was larger in Scots pine in this comparison. A comparison of silver birch and Norway spruce showed that silver birch leaf litter was significantly richer in N, Mg and K than was that of Norway spruce. It also had higher concentrations of water solubles and less lignin.

A crude generalization would state that silver birch leaf litter generally has higher nutrient levels than needle litter of Norway spruce. Spruce, in turn, is richer than lodgepole pine, which is richer than Scots pine.

Lignin levels increased in the order: Scots pine>silver birch>Norway spruce>lodgepole pine. Although lignin levels could not be related to any other chemical component across litter species, it appeared that the concentrations of water solubles, ethanol solubles and N were negatively related (Table 4.7; Berg and Matzner 1997). So far, this must be regarded as only an empirical finding.

Table 4.7. Correlations for the relationships between concentrations of N and water soluble and ethanol-soluble fractions in the newly shed litter of some conifers. (Berg and Matzner 1997)

Species	n	R	p
Water-soluble substances *vs.* N concentration			
Scots pine	10	-0.673	<0.05
Norway spruce	30	-0.114	ns
Scots pine and Norway spruce	40	-0.706	<0.001
Scots pine, Norway spruce and lodgepole pine	47	-0.316	<0.05
Ethanol-soluble substances *vs.* N concentration			
Scots pine	10	-0.809	<0.01
Norway spruce	30	-0.710	<0.001
Scots pine and Norway spruce	40	-0.165	<0.05
Scots pine, Norway spruce and lodgepole pine	47	-0.575	<0.001

n Number of samples, p significance level, *ns* not significant.

4.3.6 How much do N and lignin concentrations vary locally and regionally?

In a study on needle litters in three adjacent stands, Berg (unpubl.) found that when N and lignin concentrations were plotted against one another, the pairs clustered by species. Thus, Scots pine, lodgepole pine and Norway spruce showed consistent patterns over a four-year period (Fig. 4.7). Extending the study further, Berg (2000b) showed that these different litter types may be arranged into different, finger-print-like groups. In another study on N and lignin concentrations, this pattern was extended and repeated for a larger region (Fig. 4.7).   Needle litter of

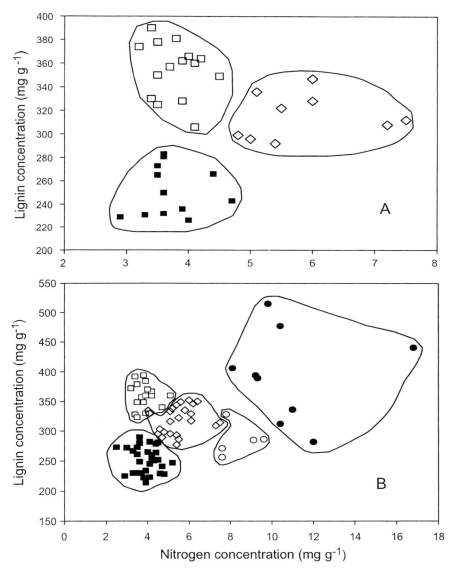

Fig. 4.7. Concentrations of lignin and N in newly shed foliar litter. **A** Foliar litter of Scots pine (■), lodgepole pine (□), and Norway spruce (◊), collected in adjacent stands for a period of four years. (B. Berg unpubl.). **B** Scots pine (■), lodgepole pine (□), Norway spruce (◊), silver birch (○), and European beech (●).   All available data. Redrawn from Berg (2000b)

Scots pine, lodgepole pine, Norway spruce, silver birch and European beech were investigated. The litter and data on chemical analyses were collected for the region of Scandinavia and for beech, it was expanded to encompass western Europe.

Scots pine needle litter formed a homogeneous group that did not overlap with the lodgepole pine or Norway spruce groups (Fig. 4.7). In this comparison, Scots pine needle litter was characterized as having simultaneously low concentrations of both N and lignin, whereas lodgepole pine litter formed another homogeneous group, with low N and high lignin concentrations. Norway spruce needles formed a group that had higher N concentrations than those of the two pine species and lignin concentrations that were in between. The birch leaves had lignin concentrations similar to those of the spruce needles and generally higher N concentrations. The leaves of European beech formed an extreme group with very high concentrations of both N and lignin.

4.4 Wood and fine root litter

The nutrient concentrations in woody litter are drastically different from those of foliage. Fine root litters, in contrast, are often rather similar to foliage in initial nutrient concentrations, but have different decay patterns. Because of the unique nature of wood and fine roots, discussion of their decay will be treated separately in Chapter 9. Here, we will briefly review the range of values observed for nutrient and organic chemical composition of wood and fine roots (Tables 4.1 and 4.8).

Nutrient concentrations are much lower in wood than in foliage litter. We may see that, for example, N concentrations may be a factor of 10 lower for the species Norway spruce, trembling aspen, silver birch and European beech (Tables 4.4 and 4.8). Wood is largely made up of cellulose, lignin and hemicelluloses in different proportions. As a whole, the woody parts of the tree are poorer in nutrients than the photosynthesizing parts. It also appears that the levels of water solubles are lower than in the corresponding foliar litters (Tables 4.4 and 4.8).

4.5 Anthropogenic influences on initial litter composition

Human activities can dramatically influence the chemical composition of newly formed litter. These effects may be either direct or indirect. Clearly fertilization with nutrients can have an effect on the nutrient composition of litter. This is true whether the nutrients (usually including some form of N) were added as part of forest practice or because of atmospheric N deposition. Here we only examine the effects of selected human activities on initial litter composition, specifically the effects on litter chemical composition of artificial N enrichment and heavy metals deposition. The anthropogenic influences on decomposition will be discussed in Chapter 12 (see also Chap. 6).

Table 4.8. Concentrations of N, water soluble compounds and lignin in wood and fine roots from some boreal and temperate tree species

Species	N	Water soluble	Lignin
		[mg g^{-1}]	
Coniferous wood			
Norway spruce[a]	0.39	37	271
White pine[b]	0.40	15	221
Coniferous fine roots			
Norway spruce[c]	3.5	210	330
Scots pine[c]	2.5	134	273
White pine[b]	9.3	135	253
Deciduous wood			
European beech[a]	0.92	35	228
Red maple[b]	0.90	22	125
Trembling aspen[a]	0.55	39	197
Silver birch[a]	0.64	26	195
Deciduous fine roots			
Black alder[d]	12.8	nd	254
Hybrid poplar[d]	9.4	nd	262
Sugar maple[b]	16.7	196	338

nd Not determined.
[a]Staaf and Berg (1989)
[b]Aber et al. (1984)
[c]Berg et al. (1998)
[d]Camiré et al. (1991)

4.5.1 N-fertilized Scots pine and Norway spruce monocultures

Additions of N to soils have been performed to simulate N deposition. Such experiments may be done by small daily additions but also by large annual additions in direct experiments with N fertilization. The latter may be useful to interpret the effects, keeping in mind that such heavy additions should be interpreted with care. With experimental dosages as high as 100-500 kg N ha^{-1} year^{-1} it appears that most of the supplied N left the system relatively quickly, with a low percentage, in the range 9 to 30%, being recovered from the topsoil (Miller 1979; Nömmik and Möller 1981; Tamm 1999). This heavy outflow of N can be attributed in part to the fact that the fertilization technique added the full amount of N fertilizer in a period of hours. As discussed by Tamm (1999) the percentage of added N that is retained in soil depends on the magnitude of the dosage, the number of additions, and the level of saturation. Using the observations by Nömmik and Möller (1981) we can estimate that 12 to 20% may be recovered with additions in the range from 150 to 500 kg meaning that, say 20 to 60 kg ha^{-1} would remain in the soil, which corresponds to annual N deposition amounts in some areas. Long-term fertilization experiments would thus be of value in illustrating long-term deposition effects.

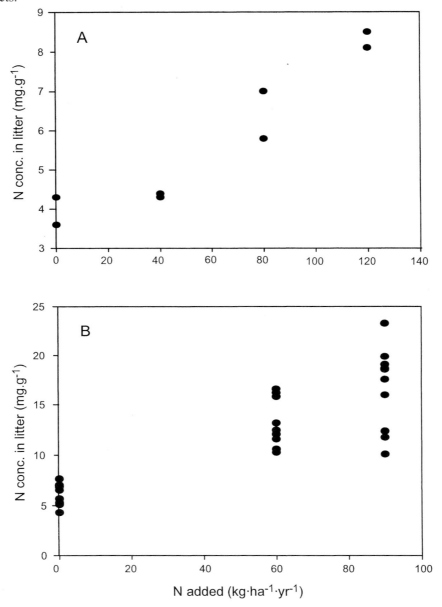

Fig. 4.8. Relationship between dosage of N fertilizer (ammonium nitrate) and concentrations of N in needle litter of: **A** Scots pine. **B** Norway spruce. (B. Berg unpubl.)

We have used data from N fertilization studies by C. O. Tamm (1991) who carried out extensive work on Scots pine and Norway spruce forests. For both species, there is a clear variation in chemical composition of needle litter between different fertilizer regimes (Table 4.9; Fig. 4.8). Tamm (1991) initially used annual doses of 50, 100 and 150 kg N ha^{-1}, later reduced to 40, 80 and 120 kg N ha^{-1}. Also with such dosages given once a year, heavy losses occurred, and the amounts retained were comparable to those of N deposition.

Table 4.9. Average concentrations of lignin, and nutrients in Norway spruce and Scots pine needle litter collected at plots with three fertilization regimes, namely control, 50, 100 and 150 kg N ha^{-1} year^{-1} and in different years. When large amounts of N are added in a single dose, as in this study, much is lost to leaching, volatilization or other mechanisms, but the additions did influence the initial lignin and nutrient concentrations of the needle litters. (B.Berg unpubl.)

Year of sample	Lig	N	P	S	K	Ca	Mg
				[mg g^{-1}]			
Norway spruce							
1983 sampling							
Control	355	4.9	0.45	0.73	0.72	17.9	0.65
100 kg N dosage	388	15.3	0.54	0.93	1.04	10.0	0.74
150 kg N dosage	402	16.8	0.55	0.98	1.27	7.7	0.76
1984 sampling							
Control	318	7.0	0.56	0.77	1.38	14.3	0.63
100 kg N dosage	347	11.7	0.64	0.84	1.78	8.4	0.78
150 kg N dosage	347	16.7	0.63	0.87	1.56	5.8	0.64
Scots pine							
1975 sampling							
Control	270	3.6	0.14	0.25	0.53	5.3	0.50
50 kg N dosage	260	4.3	0.20	0.33	0.52	5.1	0.55
100 kg N dosage	300	5.8	0.25	0.46	0.59	4.0	0.52
150 kg N dosage	380	8.5	0.30	0.49	0.85	2.9	0.38
1976 sampling							
Control	256	4.3	0.32	0.45	0.91	4.7	0.57
50 kg N dosage	251	4.4	0.30	0.43	0.98	4.6	0.67
100 kg N dosage	269	7.0	0.34	0.53	1.06	4.3	0.58
150 kg N dosage	268	8.1	0.42	0.64	1.36	4.5	0.54

Lig Lignin

The addition of N, either as fertilizer or through N deposition, will result in increased uptake by the trees and, consequently, in enhanced concentrations of N in the freshly formed litter. This has been observed by Miller and Miller (1976) and

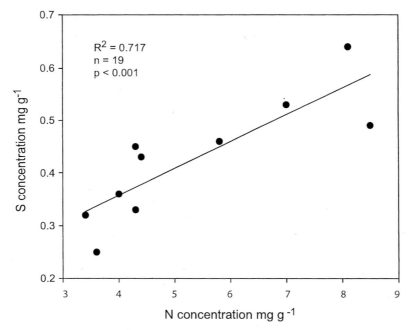

Fig. 4.9. Linear relationships between N and S concentrations in Scots pine needle litter from N-fertilized plots. (Berg and Matzner 1997)

later by Berg and Staaf (1980a). The latter, using Scots pine needle litter from a fertilization experiment (Tamm et al. 1974; Tamm 1991) found that N additions at an annual dosage of 80 kg N ha^{-1} resulted in a statistically significant increase in litter-N concentrations, whereas the dosage of 40 kg ha^{-1} year^{-1} did not have any significant effect (see Fig. 4.8). A clear relationship was seen between dosage and litter-N concentration (r=0.949, $p<0.001$, n=8). The range of the N concentrations measured over several years at one experimental site was from about 3.6 to

Table 4.10. Coefficients of determination (R^2) for linear relationships between concentrations of nutrients, water solubles, and lignin in needle litter of Scots pines subjected to different dosages of N fertilizer. A negative relationship is indicated by (-). n = 19. (Reurslag and Berg 1993)

	Component			
	N	P	S	Ca
P	0.550[**]	-	-	-
S	0.717[**]	0.932[***]	-	-
K	0.497[**]	0.901[***]	0.773[***]	-
Ca	(-)0.596[**]	-	-	-
Lig	0.517[**]	-	-	(-)0.728[***]
Wsol	0.453[**]	-	-	0.725[***]

Lig Lignin, *Wsol* water solubles, Significance levels [**] $p<0.01$, [***] $p<0.001$.

8.5 mg N g⁻¹ needle litter in control and high dosage stands, respectively. The variation in N concentration was accompanied by variation in concentrations of other nutrients as well, producing a relatively balanced nutrient composition (Fig. 4.9, Table 4.10). Thus, P, S and K concentrations showed positive linear relationships to the N concentration, whereas Ca showed a negative relationship, and there was no significant relationship found for Mg.

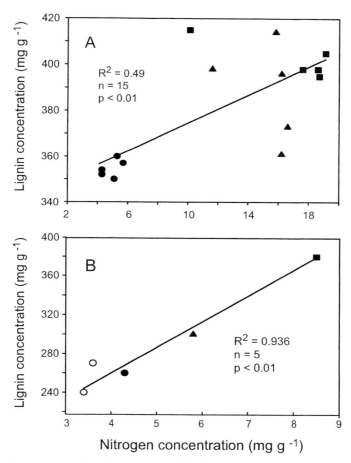

Fig. 4.10. Linear relationships between N and lignin concentrations in needle litter from N-fertilized plots with: **A** Norway spruce, and **B** Scots pine. Collections were made in 1983 (A) and 1976 (**B**). Plots were given 50 kg N ha⁻¹ year⁻¹ (●), 100 kg N ha⁻¹ year⁻¹ (▲) and 150 kg N ha⁻¹ year⁻¹ (■), control (○). (Berg and Tamm 1991; B. Berg unpubl.)

Norway spruce needle litter followed a similar pattern, although litter N had a relatively higher concentration. In general, the concentrations of N, P and S (Table 4.9) increased with dosage of N fertilizer, although the effect on the concentration of N was most pronounced. The concentrations of N in the litter appeared to be largely in proportion to the dosage of fertilizer, the range being from 4.2 mg g^{-1} in control plots to 18.3 mg g^{-1} in a high-dosage plot (Fig. 4.8). In addition, the concentration of K and Mg increased whereas Ca concentrations decreased at higher N concentrations.

It is noteworthy that concentrations of lignin also varied for both Scots pine and Norway spruce, increasing with dosage of N fertilizer. For Scots pine, the lignin concentrations increased with those of N from 270 to 380 mg g^{-1}. For Norway spruce the increase was similar, with a range of 242 to 407 mg g^{-1} (Fig. 4.10). This effect on lignin concentration seems to vary with the kind of system and appears to be indirect.

The effect could be related to a lack of B in the soil (cf. Sect. 4.3.4). The high dosage of N fertilizer may have forced the trees to grow so quickly that the supply of some essential nutrients became lacking, as the mobile pool in the soil was exhausted. Weathering apparently could not give a sufficient supply, leading to a lack of this micronutrient in the plant.

Table 4.11. Concentrations of plant nutrients and heavy metals in local fresh needle litter of Scots pine sampled at six study plots in a smelter pollution gradient in northern Sweden and needle litter sampled at an unpolluted (control) site. Concentrations of Na, Al, B, Ni, Mo, Sr and Cd did not vary in this transect. (Berg et al. 1991b)

Dist	Chemical component										
[Km]	[mg g^{-1}]								[μg g^{-1}]		
	N	P	S	K	Ca	Mg	Mn	Fe	Zn	Cu	Pb
2.5	3.78	0.26	0.99	1.43	5.23	0.47	0.79	0.38	0.25	0.100	311
3	3.73	0.24	0.73	1.01	5.70	0.53	0.83	0.36	0.19	0.068	191
7	3.25	0.19	0.49	0.70	6.11	0.46	1.26	0.14	0.11	0.019	44
9	3.71	0.26	0.50	1.08	4.65	0.56	1.10	0.27	0.11	0.012	34
13	3.66	0.25	0.53	1.23	5.65	0.66	1.43	0.12	0.08	0.009	22
30	4.40	0.22	0.51	0.98	5.70	0.67	1.21	0.11	0.07	0.006	12
Control	4.80	0.35	0.41	1.20	5.26	0.49	1.35	0.06	0.05	0.002	1

Dist Distance from smelter

4.5.2 Heavy metal pollution and initial litter chemical composition

Heavy metals can be taken up by plants through their roots or accumulate on their leaf surfaces from atmospheric deposition. In sufficient concentrations, these metals can cause a slowing of decomposition, presumably due to toxicity towards the microbial community.

Berg et al. (1991b) studied this by collecting fresh Scots pine needle litter along a transect of increasing distance from a smelter in northern Sweden. The chemical composition of needle litter collected at each site at abscission varied with the distance from the smelter (Table 4.11). A significant positive relationship ($p<0.05$) was found between the distances from the smelter and Mg concentrations in the fresh litter. The same tendency was also observed for Mn (Fig. 4.11) meaning that these concentrations increase with the distance from the smelter. Of the pollutants, Pb and Zn concentrations showed strong decreases with distance ($p<0.01$). The same trend was noted for Fe and Cu ($p<0.05$; Fig. 4.11) and, although less marked, for S and Cd ($p<0.1$, Fig. 4.11). The concentrations of organic compounds, on the other hand, seemed largely unaffected. The completely unpolluted litter had somewhat lower lignin and higher N and P concentrations than the locally collected needles as well as very low concentrations of heavy metals.

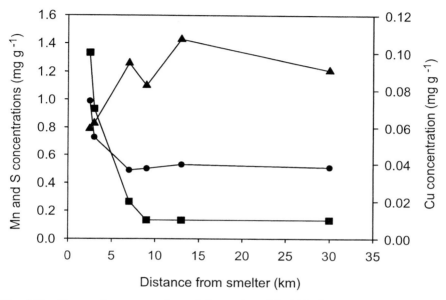

Fig. 4.11. Changes in concentrations of the nutrients Mn (▲), Cu (■), and S (●) in Scots pine needle litter collected at different distance from a smelter. (B. Berg unpubl.)

5 Changes in substrate composition during decomposition

5.1 Introductory comments

There are two principal approaches to studies of the chemical changes in litter during decomposition, namely to follow the changes in: (1) organic chemical composition, and (2) inorganic nutrient composition. We will discuss both.

The decomposition of litter organic components by microbes is selective. Thus, there is a pattern in litter chemical changes over the course of decomposition. This common basic pattern may be modified as a result of the initial chemical composition of a given litter type. The pattern discussed here is based on boreal forest systems, but probably has a wider generality. For example, even in such a different system as a chaparral, decomposition follows a pattern similar to that in boreal forests (Schlesinger and Hasey 1981).

Studies of changes in the chemical components of decomposing litter are uneven, with lignin having received much of the attention. Still, for Scots pine needle litter, more detailed descriptions have been made, including hemicelluloses, and different fractions of solubles (Berg et al. 1982a). There are also studies on Scots pine needle litter covering the decomposition process from litter fall to a close-to-humus stage. In this chapter we describe detailed chemical changes for Scots pine as a case study. We also present data from other boreal species. For K, N and lignin, specific syntheses have been published and these studies are reviewed.

The dynamics of nutrients in litter decomposition has been studied often due to its relationship to nutrient cycling in ecosystems (O'Neill et al. 1975; Anderson and Macfadyen 1976). Several such studies on nutrient dynamics also deal with chemical composition changes during decomposition (Dwyer and Merriam 1983; Berg et al. 1987; Dziadowiec 1987), and the dynamics of the major plant nutrients (Berg and Staaf 1980a,b, 1987; Blair 1988a,b; Rashid and Schaefer 1988). There are few studies, though, that describe nutrient dynamics covering the whole process from newly shed decomposing litter until the humus stage.

The limit between the concepts of "nutrients" and "heavy metals" is not clear in the literature. The general view is that nutrients are inorganic elements that are essential for life processes, whereas heavy metals are inorganic elements that, in sufficiently high concentrations, can be damaging to life processes. Although heavy metals can be anthropogenic pollutants, they also occur naturally. In this chapter, we treat heavy metals as nutrients in basically unpolluted systems and discuss their dynamics in a fundamental, natural stage. The role of heavy metals as pollutants is addressed in Chapter 12.

5.2 Organic-chemical changes during litter decomposition

Litter decomposition is normally studied under different climatic conditions and in different ecosystems, with the result that even identical litter would have different decomposition rates, and thus different concentration changes for a given component. To overcome such problems in part and to make studies more compatible in terms of chemical components, we compare litter chemical changes as a function of cumulative litter mass loss rather than of time.

5.2.1 A case study on Scots pine needle litter

Water soluble fraction

The water soluble fraction is chemically complex, resulting in a far from homogeneous substrate, and the microbial degradability of single components may vary greatly. Generally, for freshly formed foliar litter, this fraction contains high levels of simple sugars, lower fatty acids, and protein remains such as amino acids, and peptides. Most of these components can easily be taken up by microorganisms and metabolized. As a result, the water-soluble fraction should decompose rather quickly and its concentration should decrease (Fig. 5.1). For Scots pine needle litter in a boreal system, the concentration was found to decrease from ca. 92 to 57 mg g^{-1} in about a year, whereas for the sub-fraction of simple sugars and glycosides, the decrease went from 31 mg g^{-1} to undetectable amounts in the same period. Clearly, there are large differences among rates of disappearance among the compounds included in the water-soluble fraction. For some deciduous species that have been investigated, the rate of decrease of water solubles may be even more drastic (Table 5.1), and for silver birch leaf litter, the decrease in 1 year was from 321 to 40 mg g^{-1}. For Norway spruce needles, considered to start decomposing when still attached to the tree, the decrease is of a similar magnitude as for Scots pine and lodgepole pine (Table 5.1). Leaching may also play a role in the decrease in the concentrations of water solubles.

Soluble substances are formed during the decomposition of polymer compounds such as holocellulose and lignin and thus a low level of water solubles is found throughout the decomposition process. In fact, even such a rapidly decomposed compound as glucose has been found in concentrations of up to 1% in Scots pine needle litter that has been decomposing in the field for up to 5 years (Berg et al. 1982a).

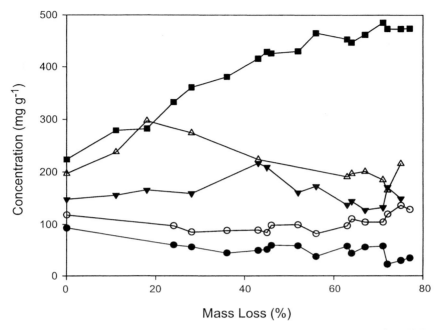

Fig. 5.1. Changes in concentrations of water solubles (●), ethanol solubles (○), cellulose (Δ), hemicelluloses (▼) and lignin (■) in decomposing Scots pine needle litter. B. Berg (unpubl.)

Ethanol soluble fraction

This fraction includes rather small molecules that are not water soluble. These are often found to be soluble in either ethanol or acetone, solvents that extract both lower phenolics and higher fatty acids. This fraction sometimes appears to contain compounds that suppress microbial growth (Berg et al. 1980) as seen for single fungal species. Furthermore, mixed microbial cultures have been found to degrade the ethanol soluble compounds more slowly than water solubles. At low pH, negatively charged stable compounds can be leached out of the litter, and the concentration of this fraction will thus decrease, although more slowly than that of water solubles. The individual components of this kind of fraction have not been determined (Chap. 4) and their degradability is thus not known.

Although the concentration of this fraction decreases as litter decomposes, new compounds will be added as degradation products from components such as lignin, will have limited solubility in water and their concentrations often remain high even after some years of decomposition (Table 5.1). For Scots pine, the concentration of ethanol solubles after 3-5 years of decomposition could be of a similar magnitude as in the initial litter. For example, in one study of Scots pine, the initial concentration of ethanol solubles was 120 mg g^{-1}. After 4 years of decomposition, the concentration had changed to 126 mg g^{-1}. The same was seen for lodgepole pine, silver birch, and grey alder.

Cellulose

We showed (Chap. 2) that in the early stage of decomposition, the insoluble components cellulose, the hemicelluloses and lignin were degraded at different rates. Somewhat later in the decay process, the degradation rates of these components became similar (Figs. 2.3, 2.4). The initially different decay rates mean that the concentrations of these components also change at different rates. Thus, the concentration of cellulose decreases slowly (Fig. 5.1). In one case with Scots pine needles, the concentration decreased slowly and at about 70% mass loss it was similar to the initial value (Berg et al. 1982a).

In later stages of decay when lignin dominates the degradation (Fig. 2.2), the degradation rates of the insoluble components become small and similar. Thus, during this late stage, the concentrations of cellulose remain rather constant (Fig. 5.1; see below).

Table 5.1. Long-term organic chemical changes in some different decomposing litter types expressed as initial concentrations and as the concentrations when the given mass-loss level was reached

Species	Wsol [mg g^{-1}]		Esol [mg g^{-1}]		Holocell [mg g^{-1}]		Lignin [mg g^{-1}]		Final m.l. [%]
	init	fin	init	fin	init	fin	init	fin	
Needle litter									
Scots pine[a]	92	34	120	126	347	92	223	472	77.1
Lodgepole pine[b]	109	44	42	53			366	482	75.3
Norway spruce[c]	114	38	60	31			344	516	51.3
Can. hemlock[d]	181	53	177	36	396	234	206	226	45.1
White pine[d]	162	18	166	46	447	219	225	185	53.2
Leaf litter									
Silver birch[b]	321	40	57	43			263	506	65.4
Grey alder[b,c]	264	33	39	36			264	475	55.5
Red oak[d]	210	55	90	15	452	171	248	213	54.6
Sugar maple[d]	336	41	112	12	431	94	121	99	75.4

Wsol Water solubles, *Esol* ethanol solubles, *Holocell* holocellulose, *fin* final, *init* initial, *m.l.* mass loss.
[a]Berg et al. (1982a).
[b]Berg and Ekbohm (1991).
[c]Berg et al. (1991a).
[d]Aber et al. (1984).

Hemicelluloses

The most common hemicelluloses decompose in a similar fashion in litter. For the most part, they behave like cellulose, although they may have different positions in the decomposing fibers. This means that the concentrations of, for example, xylans, mannans, arabinans, and galactans decrease in the early stage. Their concentrations in late, lignin-regulated, stages of decomposition become constant when compared to each other. Considering the structure and complexity of the hemicelluloses, they could be regarded as a group. When considered together, the ratio between hemicelluloses and cellulose became constant in the late stages (Fig. 5.1).

Lignin

Lignin is not a very clear concept either in fresh or decomposing litter. Lignin is often defined on the basis of proximate analytical methods rather than purely chemical criteria. When applied to freshly fallen litter or undecomposed plant materials, the proximate methods yield results close to what would be chemically defined as lignin. However, in decomposing plant litter, the lignin is modified by the humification processes, including condensation reactions, and by partial degradation by microorganisms. The formation of these decay products, which are included in the lignin fraction, may raise arguments about the extent to which true lignin is measured in decomposing litter. In addition, the lignin determined by gravimetric methods will contain, among other things, chitin from fungal mycelium and an inorganic fraction (ash) that can be of a considerable magnitude. Although this fraction for Scots pine lignin is about 1% of the total litter mass, it may amount to some percent in the gravimetric lignin analysis. The ash content of newly shed deciduous litter can be much higher, going above 10% in some cases. Furthermore, ash concentrations may increase during decay. This ash fraction should be considered when reporting lignin contents of decomposing litter.

We contend that although what is determined is not true lignin, it is a chemical fraction with similarities to true lignin. It is important to note that even native (true) lignin is highly variable among species and even within species. Thus even native lignin cannot be described with the same chemical precision as cellulose or many other plant polymers. Lignin is not a well-defined compound when it is produced and it remains a poorly defined compound as it is decomposing. The nomenclature for lignin that has been modified during decomposition is still in question. A term like "lignins" is sometimes used and even the misleading term "acid-insoluble substance" is seen in the literature. A recent suggestion is "non-hydrolyzed remains" (NHR, Faituri 2002). A term like "Lignin in Decomposing Litter" (LDL) would not be correct either, but would indicate more clearly the nature of this complex analytical fraction. It should be pointed out that although the terminology is misleading, the gravimetric "lignin" or "LDL" fraction that contains recalcitrant, "non-hydrolyzable remains" is today as important to litter decomposition studies as is a more defined lignin.

In the course of decomposition, when the more easily degradable compounds are decomposed, lignin remains relatively intact for a long time. This means that the litter becomes enriched in lignin or lignins or LDL (Fig. 5.1) and its concentration increases. A number of studies have shown that the concentration may reach even above 50% (Table 5.1). At a certain stage when the open structures of the more degradable holocellulose are decomposed, the remaining fiber will have lignin and its modified products as a protective barrier, and when holocellulose and lignin are degraded at similar rates the relative proportions of both holocellulose and lignin remain nearly constant.

5.2.2 Other species

The changes in chemical composition during decay have been studied in other species than Scots pine including Norway spruce, lodgepole pine, silver birch, European beech, sugar maple, white pine, Canadian hemlock, red oak, white oak and bigtooth aspen. The basic patterns as regards water solubles, holocellulose, and lignin are similar. There are initially large decreases in the concentration of water solubles followed by a decrease in holocellulose concentration and an increase in the concentration of lignin. In the cases investigated, the lignin fraction approaches a maximum of 45-55% (Fig. 5.2). One exception to the generally increasing lignin concentrations was found for leaf litter of European beech (Rutigliano et al. 1996), for which there was a clear decrease in concentration. This litter was rich in lignin and N and was incubated on an N-rich humus layer (Chaps. 2, 3, 6).

5.2.3 Relationships between holocellulose and lignin

Attempts have been made to describe how the concentrations of cellulose and hemicellulose change during decomposition as compared to lignin (Chap. 2). The concentration of holocellulose decreases and that of lignin increases until a level is reached at which the proportions remain constant. Two such quotients are:

HLQ = holocellulose / (lignin + holocellulose)   Berg et al. (1984)

and

LCI= lignin /(lignin + holocellulose)   Melillo et al. (1989)

The former quotient approaches asymptotically a minimum value, which may be different for different litter types (Fig. 5.3). For example, Berg et al. (1984) found a clear difference between the HLQ values for Scots pine and silver birch.

In an attempt to give a more general view on the concept of LCI, Melillo et al. (1989) introduced the concept "biological filter", and compared the development of the LCI quotient during decomposition, to the quotients found in different humus types (Table 5.2). In their study, the quotients for decomposing red pine needle litter ended at 0.67, and the values for the corresponding humus were 0.72 in the forest floor and 0.73 in the mineral soil.

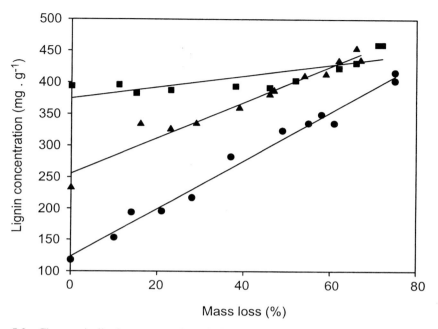

Fig. 5.2. Changes in lignin concentrations during decomposition of needle litter of Scots pine (▲), lodgepole pine (■), and sugar maple (●) with different initial lignin concentrations. Lignin concentration is plotted *vs* litter mass loss. (Berg et al. 1997)

5.3 Nutrient and heavy metal concentrations during decay

5.3.1 Changes in concentrations of elements in decomposing litter

Three principal patterns of nutrient concentrations have been observed in litter during decay, based on the release and accumulation of nutrients. Some nutrients are released from the litter at a rate that is low and proportional to overall mass loss, resulting in a linear concentration increase with cumulative mass loss. Other nutrients may be readily leached from the litter and disappear faster than the litter mass as a whole, resulting in linear or curved negative relationships to mass loss. Finally, some nutrients are strongly retained within or even imported into the litter-microbe complex, resulting in an exponential increase in concentration *vs* litter mass loss. These latter nutrients can increase in both concentration and amount during decay.

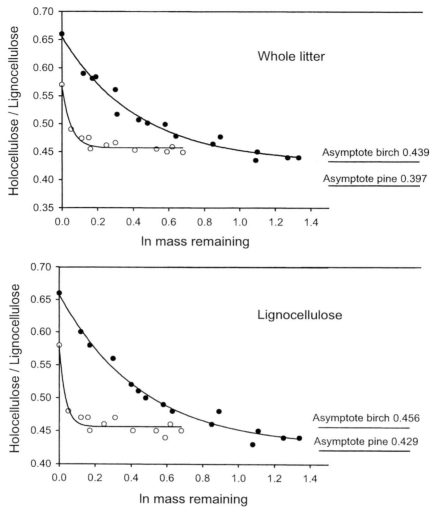

Fig. 5.3. A fitted nonlinear model gives the decrease in the quotient holocellulose/(lignin + holocellulose (*HLQ*) as a function of the natural log of whole litter mass (upper figure) and mass of lignin plus holocellulose (lower figure) remaining from Scots pine needle litter and leaves of silver birch. Also given are the asymptotic values that they approach. The litter was incubated in a nutrient-poor Scots pine forest in central Sweden. Redrawn from Berg et al. (1984)

Scots pine

The leaching of most nutrients from Scots pine litter is generally low which means that their loss from the litter is more closely related to microbial decomposition processes than to initial leaching. The relationships presented here are thought to

Table 5.2. Lignocellulose index (*LCI*) values from forest soils in northeastern United States. Modified from Melillo et al. (1989)

Species / location	Horizon	LCI
White pine stand / New Hampshire (USA)[a]	Oi	0.50
	Oe	0.66
	Oa	0.73
	A	0.78
	B	0.76
Spruce / Maine (USA)[a]	Oi	0.52
	Oe	0.73
Hardwood-spruce mix / Maine (USA)[b]	Oi	0.45
	Oe	0.63
	Oa	0.78

[a]Waksman et al. (1928).
[b]Waksman and Reuszer (1932).

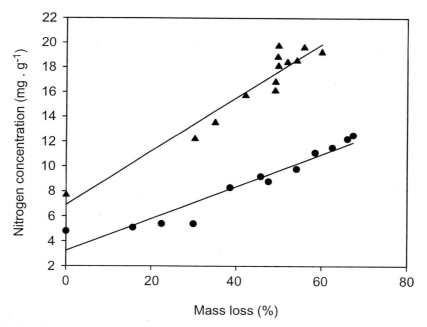

Fig. 5.4. The increase of N concentration in decomposing litter, as compared to litter mass loss. The linear relationship, which is empirical, appears to have a generality. Still the relationship is dependent on the sampling intensity. The Scots pine needle litter (●) having an initially slower decomposition than silver birch leaf litter (▲)gives a clearer linear relationship. Note the large mass loss for birch litter in the first sampling, indicating heavy leaching (B. Berg unpubl.)

be representative for pine litter in a common boreal forest type. To obtain a standard for comparing nutrient dynamics among litter types independently of decay rates we have plotted nutrient concentrations relative to cumulative litter mass loss rather than relative to time.

Nitrogen. The concentration of N in litter increases during decomposition. This increase may be described versus time since incubation, or as a function of litter mass loss, in which case we regard the decomposition process of litter as a driving force for the change in N concentration. In this latter case, a positive linear relationship results (Fig. 5.4), allowing comparisons between studies, sites and experimental treatments. This kind of linear relationship is empirical and has not been explained although it normally has R^2 values well above 0.9.

For boreal Scots pine litter, the N concentration may increase about 3-fold during decomposition, starting with ca. 4 mg g^{-1} and increasing up to ca.12 mg g^{-1} (see the special study below).

Phosphorus. The concentration of P in litter increases during decomposition, in a manner very similar to that of N. As is the case for N, this relationship can be described as a positive linear function of litter mass loss (Fig. 5.5). In addition, for P, the linear relationship is empirical. For Scots pine, a 4-fold increase from ca. 0.2 to 0.8 mg g^{-1} has been recorded (Staaf and Berg 1982).

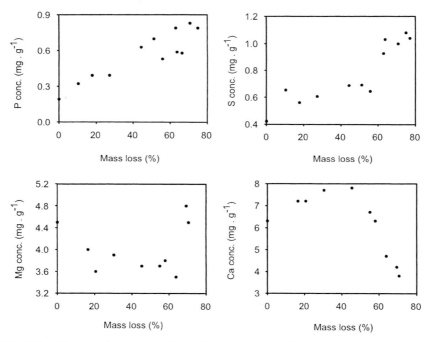

Fig. 5.5. Changes in concentration of phosphorus, sulfur, calcium and magnesium in decomposing Scots pine needle litter (B. Berg unpubl.)

Sulfur. As was the case for N and P, the concentration of S in litter increases linearly during decomposition with respect to accumulated litter mass loss. Also, in this case, the positive linear relationship is empirical (Fig. 5.5) and for Scots pine an increase from 0.4 to 1.0 mg g^{-1} has been recorded (Staaf and Berg 1982).

Potassium. As K is highly soluble, a proportion of the total is leached very soon after the litter has fallen. Normally by the first sampling in decomposition studies, a large reduction in concentration to a minimum is seen, after which the concentration starts to increase again (Fig. 5.7). Due to the high mobility, rather rapid and large changes in K concentration may take place thereby often creating uneven concentration graphs that may change considerably among studies. See also the special study below.

Calcium. A characteristic of the Ca concentration dynamics during decomposition is a peak in concentration followed by a decrease. The turning point corresponds closely to the point in decay at which net lignin degradation begins (B. Berg and C. McClaugherty unpubl.; Fig. 5.5).

Magnesium. The concentration of Mg decreases slowly without the fast leaching that was seen for K. However, as for K, the decrease is interrupted and an increase is observed. As a simplification, the graph may be described as a positive X^2 graph (Fig. 5.5).

Other nutrients/heavy metals. There appear to be too few studies on the other metals/elements to allow us to suggest that the graphs presented below are generally applicable. Still, a group of them have similar patterns with clear trends. Studies indicate that the concentrations of most heavy metals increase as the litter decomposes, even up to around 80% mass loss. The nutrients/heavy metals Cu, Pb, Fe, Zn, (Fig. 5.6) as well as Ba, Sr and Al all have a pattern of increasing concentrations during litter decomposition, essentially following an exponential graph. In all cases their concentrations increase faster than just a conservation of the existing amount would suggest. Thus, it is likely that an import takes place. Iron and Pb are known to be relatively immobile over a wide range of soil acidity (Bergkvist 1986) and are characterized by high, exponential concentration increases. In a case study for Al, the increase was from initially 280 µg g^{-1} to ca. 900 at ca. 65% mass loss. For Pb, the corresponding figures were 2.5 and 25 µg g^{-1}, for Cu 1.4 and 5, for Fe 55 and 600 µg g^{-1}, for Ba 4 and 28 µg g^{-1}, and for Sr there was an increase from ca. 5 to ca. 10 µg g^{-1}. The concentration of Cd increased from ca. 0.1 µg g^{-1} to ca. 0.4 at 65% mass loss.

At least two of the nutrients/heavy metals, Mn and Cd, show an increasing solubility and mobility with decreasing pH. They are thus often leached from litter. However, such a property is not independent of the microbial population, and at low concentrations when the metals are not in excess, a low pH does not necessarily mean a high net mobility. The microbial population could retain or import nutrients in the decomposing litter-microbe complex. Nevertheless, for Scots pine, the typical pattern for Mn was a concentration decrease and at a rate that was in proportion to litter mass loss. For other species such as silver birch the pattern of heavy metal dynamics was similar to those for Scots pine (B. Berg unpubl.).

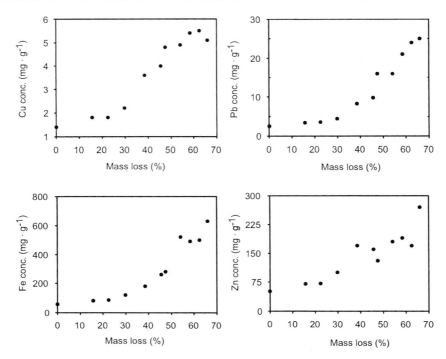

Fig. 5.6. Changes in concentrations of the heavy metals copper, lead, iron and zinc with mass loss in decomposing Scots pine needle litter (B. Berg unpubl.)

5.4 Special studies on K, N and lignin dynamics

5.4.1 K concentration dynamics

Potassium is an essential plant nutrient, and is one of the principal nutrients, along with N and P, that are routinely included in fertilizers. Potassium is one of the most mobile elements in ecosystems and its presence in dead plant tissues is highly variable. By using available data for foliar litter, Laskowski et al. (1995) found a range of 0.31-15.64 mg g^{-1} in initial concentrations in newly shed litter, thus with a range-factor of about 50. In their inventory of 139 sets of data for decomposing litter in field studies, they could systemize the dynamics of K concentration changes with decomposition, using data from boreal and temperate forest systems from mainly northern Europe and North America. The litter types investigated (mainly Scots pine, lodgepole pine, Norway spruce, silver birch, sugar maple and white pine) covered the most common litter types found in forests of northern Europe and represented some major North American species.

In an analysis using 90 sets of data, Laskowski et al. (1995) found that each set could be divided into two phases as regards K dynamics during decomposition

(Fig. 5.7). They distinguished (1) an initial phase with a fast change in concentration and, (2) a later phase with either a slow increase or stabilization in concentration. Usually, the concentration reached its lowest level (C_{min}) on the first sampling occasion, after a few months' incubation in the field. Depending on the litter type and length of incubation, the accumulated litter mass loss at the first sampling varied within a range from 2.5% up to about 40%.

These two phases were not observed for litter types with extreme K concentrations. Thus, for litters with exceptionally low initial K levels (about 0.3 mg g^{-1}) such as Norway spruce needles, an increase in concentration was observed from the very start of the incubation. On the other hand, as discussed in Chapter 6, the Norway spruce litter may already be in a late stage of decomposition at litter fall. For a few litter types with very high initial values a continuous decrease in concentration was observed throughout the whole decomposition period (e.g. for some oak and birch leaf litter).

Initial phase. For the initial phase Laskowski et al. (1995) obtained a highly significant ($p<0.0001$) relationship between the initial K concentration and slope for K release, indicating that the steepness of the slope was highly dependent on the initial K concentration (Fig. 5.7) namely, the higher the initial concentration (C_0) the steeper the slope, suggesting that this could be a general model for changes in K concentrations. The negative relationship had an R^2 value of 0.449 ($n=90$) indicating that it may explain about 45% of the variability in concentration changes across species and systems.

Coniferous litter as a separate group gave a highly significant, negative relationship with an R^2_{adj} value of 0.557 ($n=80$) thus improving the relationship in spite of a narrower span of initial K concentrations. Scots pine needle litter only revealed a highly significant relationship between the initial rate and initial K concentration (C_0) ($R^2_{adj}=0.427$; $n=63$; $p<0.0001$). The studies on Scots pine needles referred to decomposition in Scots pine forests, and the relationship they obtained explained as much as 43% of the change in concentration in spite of the fact that a very narrow range of initial concentrations was used as compared to the all-litter-type comparison. In addition, for lodgepole pine needle litter they found a highly significant relationship with $R^2_{adj}=0.952$ ($n=7$; $p<0.0001$). For deciduous litters they obtained a relatively high R^2 value of 0.58 ($n=10$; $p<0.02$), especially taking into account that only ten samples were analyzed. For Norway spruce litter no significant relationship was found.

For deciduous leaves as a group, the relationship between initial K concentration and slope of concentration decrease vs. time in the initial phase was more shallow than that for the coniferous litters, with slopes of -7.89 and -12.73, respectively. In a comparison of the relationships for coniferous and deciduous litters, the linear relationships differed significantly between them, both in the intercept ($p<0.05$) and in slope ($p<0.0001$). As the average initial concentration of K was lower for the coniferous litter than for deciduous, this may be misleading and actually suggests that concentrations of K in the needles drop more quickly than those in the leaves. However, in the very early stage, the deciduous leaves lost mass faster than coniferous litter (Berg and Ekbohm 1991) and such a conclusion simply cannot be drawn due to a lack of resolution. In comparison, a laboratory study suggested fast leaching of K from deciduous litter, in combination with a quick leaching of total soluble material (Bogatyrev et al. 1983). From that study it appears that in most cases a fast drop in concentration was observed, and it thus seems reasonable to conclude that mobile K was leached. This agrees with the observations of Bockheim and Leide (1986), Dziadowiec (1987) and Rapp and Leonardi (1988).

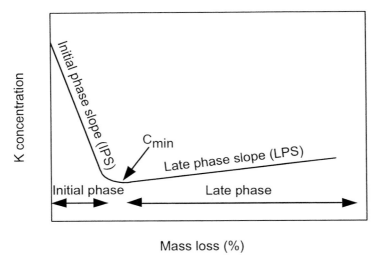

Mass loss (%)

Fig. 5.7. Overview graph for changes in K concentration with accumulated mass loss. Two phases in K concentration changes are identified. The "*initial phase*" has a rapid release of K. The "*late phase*" is the consecutive, equally clearly distinguished phase with much slower concentration changes. The initial concentration of K has been denominated C_0 and the minimum concentration for K reached during its release has been called C_{min}. (Laskowski et al. 1995)

For some litter types with exceptionally low initial K concentrations (e.g. Norway spruce litter) they found an increase in concentration from the very start of the incubations, which may mean that leaching was either very low or close to zero. Laskowski et al. (1995) suggest that in these cases the biological retention (buildup of K in the microbial biomass) overbalanced any leaching process.

Minimum concentrations.   There were significant between-species differences for minimum concentrations (C_{min}) of K in decomposing litter ($p<0.0001$; Table 5.3). The lowest level was observed for Scots pine needle litter with an average of 0.48 mg g^{-1}, and significantly higher values were found for the litter of Norway spruce (0.68 mg g^{-1}), silver birch (0.94 mg g^{-1}), oak (0.99 mg g^{-1}) and mixed pine-beech leaf litter (1.52 mg g^{-1}). The latter value was the highest average C_{min} and differed significantly from that of lodgepole pine litter (0.56 mg g^{-1}; Table 5.3).

Late phase.   At the later stages of decomposition, the concentrations of K approached a final level that appeared to be similar for the different litter types, irrespective of litter species and more similar than the initial concentrations. While the range of initial K concentrations (C_0 values) for the whole data set was as large as 0.31 to 15.64 mg g^{-1}, with a range factor of ca. 50, the final range was only 0.37 to 1.78 mg·g^{-1}, with a range factor of ca. 5. These final values were dependent on the accumulated mass loss at the last sampling occasion for each set, and in order to obtain more comparable values, Laskowski et al. (1995) introduced a value for the concentration of K at 60% mass loss (C_{60}), which was used as a standardized value for comparisons. At 60% mass loss they considered the litter to be in a humus-near stage.

The standardized C_{60} value gave an even narrower range in late-phase concentrations with 0.78 and 1.43 mg·g^{-1} as extreme C_{60} values, thus a range factor of ca. 2 (Table 5.3). In contrast to the initial concentrations (C_0) and the C_{min} values, there were no significant differences in K concentrations at C_{60} between litter types.

In summary, the initial concentrations were highly significantly different among the litter types. After some decomposition the differences were smaller, but still significant, and finally at C_{60} no difference was detected. Thus, the initially large and statistically significant differences in average initial K concentrations between coniferous and deciduous litters (average value 1.03 vs. 4.54 mg·g^{-1}, respectively) decreased in the late phase and at C_{60} they practically disappeared (1.06 vs. 1.19 mg·g^{-1}, respectively; Table 5.3).

5.4.2. N concentration dynamics over a climatic transect

Repeatability of N concentration increase. That N concentrations increase in decomposing litter is widely known. When the increase in N concentration is related to time since incubation, the result is a curve with an asymptotic appearance. We have related the increasing N concentrations to litter mass loss (cf. above) for several litter types, resulting in a linear increase, possibly until the limit value for decomposition is reached (Aber and Melillo 1982; Berg et al. 1997; Fig. 5.4). Such a linear increase has been found for many species including foliar litter of Scots pine and Norway spruce. Deciduous litter species, such as silver birch, also give linear relationships, but much mass is lost initially, resulting in a fast increase in N in proportion to mass loss (Fig. 5.4). For Scots pine, the increase in concentration is linear from an initial 4 mg g^{-1} up to 12 to 13 mg g^{-1} N at about 75% mass loss (Fig. 5.8). Still, the reasons for the straight-line relationships are far from clear, given the simultaneous in- and outflows of N during the decomposition process (Berg 1988). This linear relationship is an empirical finding and for systems such as coniferous foliar litter, the relationship appears to be highly significant (Fig. 5.9).

Table 5.3. Potassium concentrations in decomposing litters, including both green and brown litter. C_0 = initial concentration, C_{min} = minimum concentration, C_{60} = concentration at 60% mass loss. Standard errors in parentheses, n = number of incubations. (Laskowski et al. 1995)

Litter type	C_0	C_{min}	C_{60}	n
All litters	1.48(0.15)	0.57(0.02)	1.08(0.03)	139
Coniferous litters	1.03(0.06)	0.51(0.02)	1.06(0.02)	121
Scots pine needles	0.88(0.05)	0.48(0.02)	1.06(0.03)	97
Lodgepole pine needles	1.03(0.47)	0.56(0.06)	1.17(0.19)	7
Norway spruce needles	1.90(0.20)	0.68(0.04)	0.99(0.04)	17
Mixed pine – beech litter	4.10(0.54)	1.52(0.15)	1.18(0.46)	2
Deciduous leaves	4.54(0.87)	0.90(0.08)	1.19(0.12)	16
Silver birch leaves	5.01(0.49)	0.94(0.11)	1.29(0.23)	8
Grey alder leaves	8.26(7.39)	0.69(0.18)	0.97(0.32)	2
Common oak leaves	2.13(0.69)	0.99(0.17)	1.14(0.09)	3
European maple leaves	4.19	1.34	1.43	1
European beech leaves	1.67	0.69	1.16	1
Mixed oak – hornbeam leaves	3.72	0.52	0.78	1

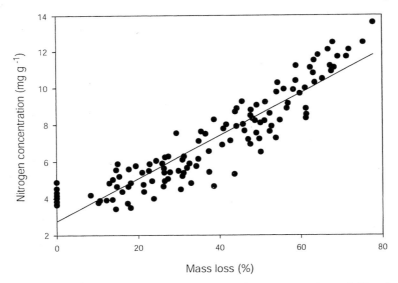

Fig. 5.8. Overall relationship between increasing concentrations of N and accumulated mass loss for decomposing Scots pine needle litter. Incubations were made at one site, a nutrient-poor Scots pine forest. Data are pooled from 14 sets, each representing an incubation of local litter from a different year. (Berg et al. 1997)

Berg et al. (1997) compared the linear relationship for accumulated mass loss vs. N concentrations among several sets of decomposing Scots pine needle litter in one system (Fig. 5.8). They called the slope of the relationship the Nitrogen Concentration Increase Rate (NCIR). The litter was naturally produced from the Scots pine monocultural system, and the variation in initial N concentration was the observed annual variation. For one site, the relative increase rates in N concentration showed significant linear relationships to litter mass loss for individual sets of litter, as well as for the average combined from all 14 sets of litter (Table 5.4). The NCIR values in this comparison had an average of 0.1109, and the slopes ranged between 0.055 and 0.129 (SE = 0.0047) indicating that for a given litter type and system the variation in NCIR was somewhat limited.

In a comparison of NCIR values for lodgepole pine needle litter, the slopes of five lodgepole pine data sets gave similar results with an average slope of 0.1151 and a standard error of 0.0060 (Table 5.4). For needle litter of Norway spruce, the average slope was similar to that of the pine litters and reasonably consistent among four sets of litter (Table 5.4). The conclusion may be that although the increase in N concentration during decomposition is empirical, it is consistently observed. When the three different studies are compared, we see that the two pine species, having similar N levels, increase in parallel (Table 5.4), whereas the Norway spruce litters, as a result of initially higher N concentrations, form a group above the pine litter, while having similar slopes (Fig. 5.9).

Table 5.4. Overall *Nitrogen Concentration Increase Rates (NCIR)* for local needle litter of three coniferous tree species incubated at three sites. Comparisons are made both by calculating regressions over the combined data set, and by taking the average slopes of individual data sets. Nitrogen concentrations were regressed *versus* litter mass loss. (Berg et al. 1997)

Site – species	Intercept (SE)	Slope (NCIR) (SE)	R^2_{adj}	R	n	p
Jädraås - Scots pine						
All values combined	2.941 (0.985)	0.1107 (0.0042)	0.843	0.919	131	<0.001
Average of 14 slopes		0.1109 (0.0047)			14	
Malung - Lodgepole pine						
All values combined	2.762 (1.128)	0.1171 (0.0065)	0.862	0.928	54	<0.001
Average of 5 slopes		0.1151 (0.0060)			5	
Stråsan - Norway spruce						
All values combined	4.769 (1.124)	0.1019 (0.0105)	0.638	0.799	56	<0.001
Average of 4 slopes		0.1151 (0.0031)			4	

SE standard error, *n* number of data sets.

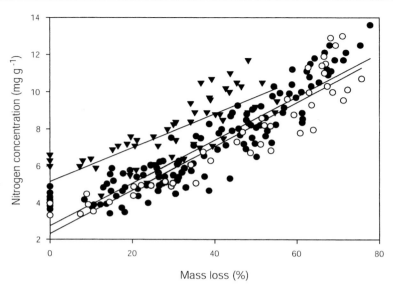

Fig. 5.9. Litter N concentration as a function of accumulated mass loss for decaying coniferous litters (●) Scots pine - 14 sets, (○) lodgepole pine - five sets, (▼)Norway spruce - four sets. Coefficients of determination (R^2) are 0.845 (n=131), 0.851 (n=54), and 0.638 (n=56), respectively. (Berg et al. 1997)

The influence of litter N concentrations. Scots pine green needles with a higher initial N concentration had a much larger NCIR than brown needles, meaning that the increase rate was larger. A similar trend was observed for decomposing Norway spruce needles. Both green needles and N-enriched needles, collected from N-fertilized plots, had higher NCIR values than the more N-poor regular brown needles. However, the difference between natural and fertilized needles was not significant (Berg et al. 1997). This property of NCIR, to increase with increasing initial N concentration, appeared to be in common for different species. Thus natural needle litter of lodgepole pine and Scots pine had similar initial N concentrations and the two species also had similar average NCIR values.

Significant relationships between initial litter N concentrations and NCIR values were found within the Scots pine species (Fig. 5.10, Table 5.5), whereas for Norway spruce needles alone, Berg et al. (1997) found no relationship. In an attempt to find a global relationship they considered all available data both from coniferous and deciduous litter and found a significant relationship, which seems to hinge on the relatively few extreme points corresponding to green and deciduous litters (Fig. 5.10). Thus, on that scale NCIR is only weakly influenced by initial N concentrations, except when considering extreme cases. However, to the extent that a relationship does occur, it is a positive one. This means that N concentrations increase to a somewhat greater extent with accumulated mass loss when initial N concentration is higher. Berg and Cortina (1995) also saw this when comparing NCIR for seven very different litter types incubated in one system (Fig. 5.11).

One mechanism for conserving N in decomposing litter could be via covalent bonding to macromolecules during the humification process. An example of this is the ammonium fixation described by Nömmik and Vahtras (1982). Thus, if the initial amount of N in the litter were higher, there would be more available for fixation, giving a higher NCIR. Such a conclusion is reasonable since Axelsson and Berg (1988) found that the N availability is limiting the rate of the process.

Table 5.5. Linear relationships describing the variation in *Nitrogen Concentration Increase Rates* (*NCIR*) as compared to the litters' initial concentration of N. The stands are natural systems. The litter species as well as their chemical composition varies between comparisons. Scots pine litter included local, fertilized and green litters. Green litter was collected on the trees well before litter fall, and experimental litter refers to litter sampled from fertilized trees. (Berg et al. 1997)

Combination	Intercept (SE)	NCIR (SE)	R^2_{adj}	R	n	$p<$
Scots pine	0.0693 (0.0221)	0.0123 (0.0012)	0.542	0.740	84	0.001
Deciduous-all available	0.1538 (0.0309)	0.0094 (0.0014)	0.860	0.938	8	0.001
Deciduous-no alder	0.0970 (0.0204)	0.0158 (0.0024)	0.880	0.949	7	0.01
All data	0.0767 (0.0367)	0.0105 (0.0009)	0.530	0.724	116	0.001

SE standard error, *n* number of data sets.

A transect of Scots pine systems. For local natural Scots pine needle litter and a unified Scots pine needle litter preparation, the relationship between NCIR and AET was investigated across a climatic transect in Sweden, with AET ranging from 380 to 520 mm. There was a highly significant relationship for Scots pine ($R^2_{adj}=0.640$, $n=31$, $p<0.001$) indicating that the N concentration will increase faster (relative to mass loss) under a warmer and wetter climate. This correlation was significant when both local and unified needle litter was used and also when using only local needle litter ($R^2_{adj}=0.517$, $n=18$, $p<0.001$).

Over a large group of litter species, and for litter collected over a broad region, the initial N concentration had minor influence as a regulating factor on NCIR. It was found (Berg et al. 1995a) that initial N concentrations in Scots pine needle litter varied over a large region and could be related to the climatic index AET. In the analysis (above) comparing NCIR to initial N levels, the pines formed a cluster (Fig. 5.10) with relatively low initial N concentrations and low NCIR values.

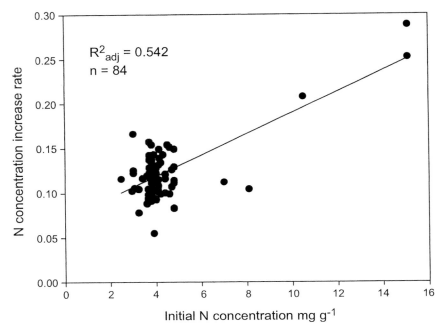

Fig. 5.10. Linear relationship between initial concentration of N in decomposing Scots pine needle litter and the slope of N concentration *vs.* accumulated mass loss *(Nitrogen Concentration Increase Rate or NCIR)*. Many individual points are obscured due to overlap. (Berg et al. 1997)

A transect of Norway spruce systems. Norway spruce litter was investigated separately (Berg et al. 1997). In a climatic transect in Sweden the NCIR values increased with increasing AET values and the relationship was highly significant (R^2_{adj}=0.534, n=14, p<0.01).

A general relationship among species. Berg et al. (1997) found that for Norway spruce, NCIR increases slightly more with increasing AET than for Scots pine litter (slopes of 0.00055 and 0.00038, respectively; Fig. 5.12), but the difference was not significant. In the comparison between NCIR and initial N concentration, there were differences between the Norway spruce litter and that of pine. They concluded that in the analysis *vs.* climate the relationships for Norway spruce and Scots pine were sufficiently similar to allow a stepwise combining of the data. First the values for Scots pine and Norway spruce were combined, resulting in a highly significant relationship with a low standard error (R^2_{adj}=0.604, n=45, p<0.001). This means that for both Scots pine and Norway spruce the climatic factor was more important for the increasing concentration of N in litter than was the species or the initial N concentration.

In a second step they combined all brown coniferous litter and obtained a highly significant linear relationship with R^2_{adj}=0.569, n=53, p<0.001; Fig. 5.12). It may be worth pointing out that the white pine needle litter data came from a more southern site with a relatively high AET but did not deviate from the general relationship.

Finally, in an analysis combining coniferous and deciduous litters, the relationship was still highly significant (R^2_{adj}=0.628, n=59, $p<0.001$). Deciduous litters departed from the pattern exhibited by coniferous litters, but the overall relationship remained significant. Thus climate, as indexed by AET, is a significant factor in affecting the rate of N concentration increase in decomposing leaf litter. As these increases were calculated on the basis of accumulated mass loss rather than time, the results mean that, after a certain mass loss, a particular litter decaying in an area with higher AET will contain more N than the same litter decaying in an area with lower AET.

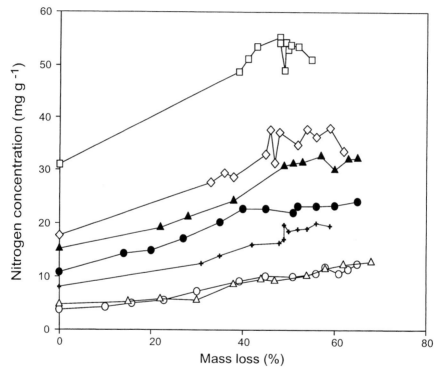

Fig. 5.11. Changes in N concentration as related to accumulated litter mass loss for seven litter types incubated in a 130-year-old Scots pine forest. Brown Scots pine needle litter (Δ), green Scots pine needles (▲), brown needles of lodgepole pine (○), green needles of lodgepole pine (●), brown leaf litter of silver birch (♦), green leaves of silver birch (◊), green leaves of gray alder (□) (Berg and Cortina 1995)

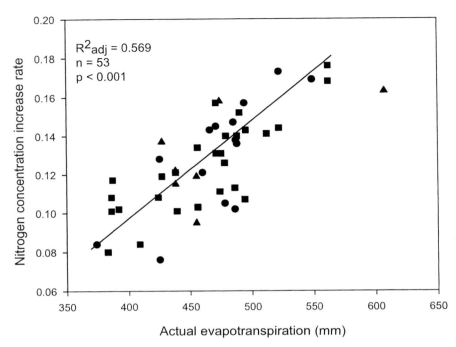

Fig. 5.12. A linear relationship between the climatic index actual evapotranspiration (AET) and *Nitrogen Concentration Increase Rate (NCIR)* in decomposing needle litter of (●) Scots pine, (■) Norway spruce, and (▲) other pines (lodgepole pine and white pine (Berg et al. 1997)

5.4.3. Lignin dynamics in a climatic transect

Repeatability of lignin concentration changes. Lignin concentration changes may be related to duration of decay, which often can be described by an asymptotic curve. The lignin concentration may, on the other hand, be related to accumulated litter mass loss, resulting in a linear increase (Fig. 5.2). Such a linear increase in lignin concentration has been found for Scots pine, lodgepole pine and Norway spruce. Deciduous litter like birch leaves give linear relationships too, but much mass is lost initially, resulting in a quick increase followed by more constant values (Berg et al. 1984). For Scots pine, the lignin concentration increases up to ca. 50% and above (Fig. 5.2) for reasons discussed earlier (Chap. 2), and in this interval, the linear relationship is highly significant (Fig. 5.13).

Berg et al. (1997) called this linear increase LCIR (Lignin Concentration Increase Rate). The LCIR during decomposition appears to be repeatable within a given stand. Native Scots pine needle litter was incubated annually for 14 consecutive years and LCIR values for the 14 sets of decomposing needle litter were compared (Table 5.6). The difference between years was the between-year variation in initial chemical composition (Berg et al. 1993e; Johansson et al. 1995; Table 4.3) and the annual variation in climate. Comparing the linear equations for lignin concentration increase among the 14 sets, the differences were found in the intercepts rather than in the slopes, the intercepts reflecting the initial lignin concentration. The average slope was 3.08 for $n=14$ (Table 5.7). The slope when using all measurement points in one linear regression was 2.92 with an R^2_{adj} of 0.894 (Fig. 5.13; Table 5.7).

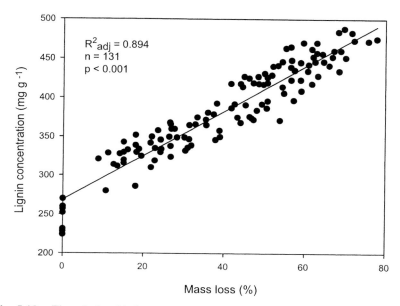

Fig. 5.13. The relationship between the increase in lignin concentration and accumulated mass loss (*Lignin Concentration Increase Rate or LCIR*) for 14 different incubations of local Scots pine needle litter at a site with Scots pine forest on nutrient-poor soil (SWECON site Jädraås). All measurement points are shown together with the common linear regression line. (Berg et al. 1997)

Two further boreal coniferous species were investigated (Berg et al. 1997), namely lodgepole pine and Norway spruce. These also showed repeatable increases in lignin concentration (LCIR). Thus, for lodgepole pine, five individual incubations at the same main site showed a low variation among slopes (average value was 1.21; $n=5$), while the slope when all the five data sets were combined became 1.24; $R^2_{adj}=0.610$; $n=55$; Table 5.7). A similar comparison for native Norway spruce litter, using data for incubated litter at one site, gave straight lines for four combined data sets of decomposing native Norway spruce needle litter, with a slope of 2.95 ($R^2_{adj}=0.841$; $n=56$). When investigating the difference among the slopes between the five individual incubations, the average slope was found to be 2.92 (Table 5.7).

Table 5.6. Linear relationships between the *Lignin Concentration Increase Rate* (*LCIR*) and litter mass loss for Scots pine needle litter incubated in its own stand (SWECON site at Jädraås). Lignin analysis was made on all samples during decomposition. (Berg et al. 1997)

Initial lignin conc. [mg g^{-1}]	Intercept	(SE)	Slope LCIR	(SE)	R^2	n
223	249.2	22.9	3.255	0.025	0.932	14
267	251.1	12.8	3.364	0.214	0.972	9
231.4	253.4	13.9	2.783	0.195	0.953	12
270	287.7	16.5	2.298	0.364	0.889	7
227.3	272.8	28.2	2.747	0.765	0.720	7
257	269.8	19.0	2.979	0.373	0.970	4
224	256.2	29.1	2.784	0.550	0.865	6
228.6	248.7	15.3	2.850	0.255	0.947	9
227.3	272.8	28.2	2.747	0.765	0.720	7
257	269.8	19.0	2.979	0.373	0.970	4
224	256.2	29.1	2.784	0.550	0.865	6
228.6	248.5	15.3	2.850	0.255	0.947	9
228.6	250.0	16.3	3.450	0.274	0.952	10
228.6	229.5	9.7	3.123	0.138	0.983	10

SE Standard error, *n* number of data points.

The influence of initial lignin concentration. Different litter types have different behaviors with respect to lignin disappearance. For litters rich in lignin, for example lodgepole pine and Norway spruce needle litter, lignin disappearance begins at or soon after litter decomposition has started (Berg and Lundmark 1987; Berg and Tamm 1991). Even in these cases, the concentrations of lignin increase as decomposition proceeds. There is, however, variation between LCIR values for different litter species collected at and incubated in their own ecosystem. At a site with monocultures of lodgepole pine and Scots pine, the litter of lodgepole pine had a lignin concentration of about 350 mg·g^{-1} and Scots pine about 290 mg g^{-1}. Both litter types had significant linear relationships between accumulated mass loss and the increase in lignin concentration, with the slopes (LCIR) being 1.24 and 2.55, respectively. The litter with initially higher lignin concentrations (lodgepole pine) had significantly lower slopes (see also Table 5.7).

In a comparison of five different data sets each for lodgepole pine and Scots pine, Berg et al. (1997) found a highly significant negative relationship between LCIR and initial lignin concentrations, with an R^2_{adj} value of 0.938 indicating that the higher the initial lignin concentration, the lower the rate of increase. This is probably due to an apparent upper limit for lignin concentration of about 50% (Fig. 5.2). It appears that there is also a more general relationship between initial lignin concentration and LCIR. Over the three pine species, Scots pine, lodgepole pine and white pine, this resulted in a highly significant negative relationship with

$p<0.001$ ($R^2_{adj}=0.471$, $n=69$). Berg et al. (1997) investigated a larger data set and examined the influence of both climate and initial lignin on the LCIR (below).

Table 5.7. Linear regressions of lignin concentration in decomposing litter vs. mass loss for needle litter of three species incubated at their own stands. The slope of the linear relationship was called *Lignin Concentration Increase Rate (LCIR)*. Comparisons of LCIR are made both by combining all values from different litter incubations into a single regression and by taking the average slopes for individual studies. (Berg et al. 1997)

Species	Intercept (SE)	LCIR (SE)	R^2_{adj}	n	$p <$
Scots pine					
All values	262.0 (20.8)	2.924 (0.089)	0.894	131	0.001
Avg. of 14 slopes		3.08 (0.091)		14	
Lodgepole pine					
All values	370.6 (25.6)	1.24 (0.134)	0.610	55	0.001
Avg of 5 slopes		1.21 (0.118)		5	
Norway spruce					
All values	362.7 (18.7)	2.95 (0.174)	0.839	56	0.001
Avg of 4 slopes		2.92		4	

SE Standard error, *n* number of data points, *Avg.* average

For Norway spruce, they investigated the relationship between initial lignin concentration and LCIR using data from an intensively studied site (Berg and Tamm 1991). The negative relationship obtained between initial lignin concentrations and LCIR was again significant. The relationship between a higher initial lignin concentration and a lower LCIR slope thus also holds for spruce. The relationship between LCIR and initial lignin concentration still held when they combined all data from coniferous litters ($p<0.001$) and gave an R^2_{adj} value of 0.313 with $n=94$.

Variation in LCIR with climate. For Scots pine, clear differences have been found in the magnitude of the LCIR between northern and southern sites in Scandinavia. In their studies, Berg et al. (1997) used a unified litter preparation over a transect ranging over most of Sweden and into northern Germany. The resulting LCIR values were then related to climate (Berg et al. 1997) using the climate index actual evapotranspiration (AET) (Meentemeyer 1978). The LCIR for Scots pine gave a highly significant positive relationship with AET with $R^2_{adj}=0.545$, and $p<0.001$ for $n=30$ (Fig. 5.14).

Combined data for Scots pine, lodgepole pine and white pine incubated all over Sweden and on Black Hawk Island (USA, see Appendix III), showed that the relationship between LCIR and AET was highly significant, and a multiple regression using both lignin and AET gave a strong fit ($R^2_{adj}=0.697$; $n=69$). Norway spruce litter shows a similar relationship between LCIR and AET ($R^2_{adj} = 0.546$; $n = 14$; $p<0.01$; Berg et al. 1997).

In an approach towards a broader relationship, they combined all available data for Scots pine, white pine, lodgepole pine and Norway spruce, and compared LCIR to AET and added initial lignin concentration (see above). For this combined set of coniferous litter, AET was a moderate predictor of LCIR but weaker than initial lignin. This regression, including both AET and lignin, gave an R^2_{adj}

of 0.697 (*n*=57) with a highly significant relationship. This means that over four coniferous species in a mainly North-European climatic transect, the climatic index AET and the initial lignin concentration could explain 70% of the variation in the increase in lignin concentration with decomposition.

Berg et al. (1997) interpreted these findings to mean that at higher AET values, more favorable climatic conditions exist for initial decomposition. With climate less limiting, the faster growing fungi would have an advantage over slower growing lignin degraders (cf. Chaps. 2 and 6). In colder climates, lignin degraders would grow relatively better as compared to those fungi degrading holocellulose. This would result in relatively more lignin being degraded at the sites with lower AET. The result would be that the higher the AET the more the lignin concentration increased per unit mass lost. In other words, for litters decomposing at sites with higher AET, the amount of lignin and its recombination products (cf. Glossary) occurring in decomposing litter at a particular percent of mass loss is greater than for the same litter at sites with lower AET values.

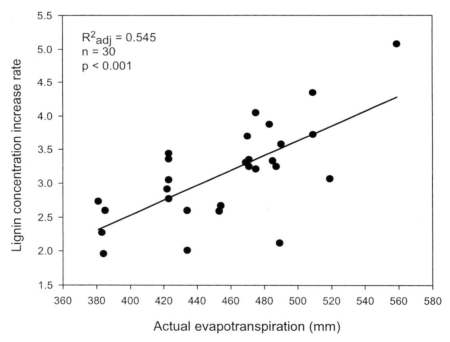

Fig. 5.14. Linear relationship between *Lignin Concentration Increase Rate (LCIR)* in unified decomposing Scots pine needle litter and the climate index *AET* in a climate transect ranging from the Arctic Circle to the northern part of the European continent. (Berg et al. 1997)

An alternative and speculative explanation may be connected to the increasing N levels in the same litter types (see above). The increase in litter N levels are also positively related to both the AET (above) and the effect of what in general terms is called the "humification process", which would be faster under warmer and wetter conditions. That would imply that N is more quickly mobilized, thus increasing the relative N concentration in the soil, allowing a faster transport to the litter. Further, the higher the N concentration, the faster the adsorption of N to lignin remains (Stevenson 1982). Such an explanation is possible since the N concentration is the limiting factor in the process of adsorption of N to decomposing litter (Axelsson and Berg 1988).

Table 5.8. Linear regressions for combined data sets for three stands. The increasing lignin concentration in decomposing litter was regressed *versus* litter mass loss using available data points for each litter type. The resulting slope is the *Lignin Concentration Increase Rate (LCIR)*. Site Jädraås is a nutrient poor stand and sites Norrliden and Stråsan are more fertile. The stands are located in monocultural systems that are not manipulated. (Berg et al. 1997)

Site	Intercept (SE)	LCIR (SE)	R^2_{adj}	R	n	$p<$
Jädraås - Scots pine needle litter						
-Local natural	262.0 (20.8)	2.946 (0.089)	0.894	0.946	131	0.001
-Exper./fertilized	263.0 (21.1)	3.236 (0.14)	0.912	0.956	53	0.001
-Green needles	249.5 (16.9)	3.16 (0.26)	0.883	0.967	12	0.001
Norrliden - Scots pine needle litter						
-Local natural	276.0 (27.9)	2.989 (0.503)	0.795	0.895	30	0.001
-Exper./fertilized	261.1 (24.2)	4.25 (0.22)	0.912	0.96	37	0.001
-Green needles	211.8 (8.6)	5.17 (0.12)	0.933	0.974	14	0.001
Stråsan -Norway spruce needles						
-Local natural	362.7 (18.7)	2.95 (0.176)	0.839	0.920	56	0.001
-Exper./fertilized	393.6 (29.1)	2.90 (0.172)	0.714	0.845	14	0.001
-Green needles	288.05 (40.2)	3.39 (0.61)	0.749	0.880	11	0.001

SE Standard error, n number of data points, *Exper/fertilized* litter from N-fertilized trees.

The influence of N concentration. Some data (Table 5.8) have indicated a higher LCIR value for nutrient-rich litter. Such an observation was judged by Berg et al. (1997) to be reasonable, since N has been observed as part of humus formation (Nömmik and Vahtras 1982). Berg et al. (1997) compared three different groups of Scots pine needle litter incubated at a nutrient-poor Scots pine site. The litter sets had different initial nutrient composition but similar initial organic-chemical composition. The brown, local natural litter had lower levels of N, whereas the needle litter originating from experimental, fertilized plots had higher levels, with the green needles having the highest concentrations. The local natural litter gave a composite (LCIR) slope of 2.95 ($n=131$) and the set of experimental Scots pine needle litter with higher initial nutrient concentrations (Berg et al. 1987) gave an average slope of 3.24 ($n=53$; Table 5.8). Finally, the set of green needles gave an LCIR value of 3.16 ($n=12$). Although small, these differences suggested that within the same litter type and the same site there may be a difference in rate of increase in lignin concentration when the nutrient concentration varies at litter fall. A similar comparison among Scots pine needle litters at a more fertile site (Norrliden; Table 5.8), with Scots pine growing on soil derived from glacial till, gave more pronounced results. In this case, the differences between groups were significant. Local, natural brown needles had an LCIR value of 2.99, whereas litter from fertilized trees gave a slope of 4.25 and the green needles gave a slope of 5.17 (Table 5.8). Needle litter of Norway spruce exhibited similar, though non-significant, trends with needles yielding higher LCIR values. Local, natural, brown needles had an LCIR of 2.95, whereas litter from fertilized trees a slope of 2.90 and the green needles gave a slope of 3.39 (Table 5.8).

There were no significant global relationships between N concentrations and LCIR among any of the data combinations analyzed. It thus appears that even though the nutrient level within a species of litter may affect the rate at which lignin concentration increases, N concentration is not a major factor.

6 Influence of chemical variation in litter on decomposition

6.1 Introduction

Chapter 5 reviewed the effects of initial litter chemical composition on the pattern of changes that occur during decay. This chapter focuses attention on the influence of changing litter quality on the decay processes, and will show that the influences of selected litter chemical components change dramatically during the process, sometimes even reversing the direction of their effect. For this purpose, the three-phase model introduced in Chapter 2 is applied to explain the effects of chemical variation, changes in mass-loss rates and decomposition patterns. Litter types that have been found to deviate from the general pattern will also be discussed.

Chemically distinct litter species or litter that has been chemically modified as a result of such things as fertilization or pollution will create different patterns during the decomposition process. Higher concentrations of the main nutrients N, P and S, may increase the initial decomposition rate and in later stages change the decomposition pattern. The case study litter (Chap. 2) showed a basic pattern, and in this chapter we will describe the effects on this basic pattern of variations in the concentrations of the main nutrients. Nutrients other than the principal ones may be important in some litters. For some litter species, notably Norway spruce needles, the patterns differ completely and initial rates may be related to nutrients such as Mn.

The decomposition process normally reaches a stage at which decomposition almost stops, or proceeds so slowly that the stage may be approximately described mathematically by an asymptote, or as a limit value for the decomposition process. For foliar litter, the limit value is normally in a range between 50 to 100% mass loss. The value is negatively related to initial litter N concentrations, which means that the more N-rich (and generally more nutrient rich) the litter is, the sooner the limit is reached and the less the litter will decompose under comparable environmental conditions. This relationship, which has been generalized for foliar litter types, is developed in this chapter.

Recent findings have indicated that raised N concentrations in litter may support the process of leaching of C compounds from that litter (Fog 1988). This leaching may begin in the early phase and continue through the remaining phases. There are extreme cases reported such as an actual disintegration of very N-rich humus that was due to a change in the microflora. This resulted in a very rapid degradation accompanied by a leaching of N-containing compounds (Guggenberger 1994). These findings will receive further attention in Chapter 11. The intention of this chapter is to demonstrate and systemize the changing effects of several chemical components on rates and patterns of litter decomposition.

6.2 A three-phase model applied to litters of different chemical composition

The decomposition dynamics in most needle and leaf litters investigated to date follow the model presented in Fig. 6.1. The model appears to be appropriate for needle litter of different pine species, as well as different types of foliar and root litter, and probably also litter from grasses and herbs. The model thus appears to have a relative generality.

6.2.1 Overview of the model

Different plant litter types have different chemical compositions when shed (Chap. 4). Some of these properties are also reflected in later stages of decomposition and higher initial concentrations of N and lignin correspond to higher concentrations of these components during the whole decomposition process (Chap. 5). To describe this we have chosen to apply and develop the three-stage model described earlier by Berg and Matzner (1997).

   For newly shed Scots pine litter with different nutrient levels, the decomposition rate was linearly related to concentrations of total N, P and S, until an accumulated mass loss of between 26 and 36% was reached (Fig. 6.2), corresponding to an early phase. The later stages are similar to those described before, and are characterized by a lignin-mediated suppression of the decomposition rate. The late stages of decomposition, and the concept of the limit value are now discussed with respect to the varying litter chemical composition. The humus-near stages are explained with respect to limit values and their relationship to the chemical composition of litter. Different litter types, however, have somewhat different patterns in relation to the model and that is discussed. We can thus see three different patterns of foliar litter decomposition, and these patterns appear to be distinct for the groups in both the early and late stages. We have given each group provisional names: "pine", "deciduous" and "spruce" and they are discussed in that order (Table 6.1).

   Even if we can set a reasonably clear border between the early and the late stages of decomposition, there is no clear border between the late stage and that which we call the "humus-near" or "limit value" stage. It is even more difficult to distinguish clearly between litter in the "humus-near" stage and true humus. Several of the functional properties, such as the effect of N on decomposition, appear to be common to different stages. Thus properties of the humus-near stage may persist into the humus stage.

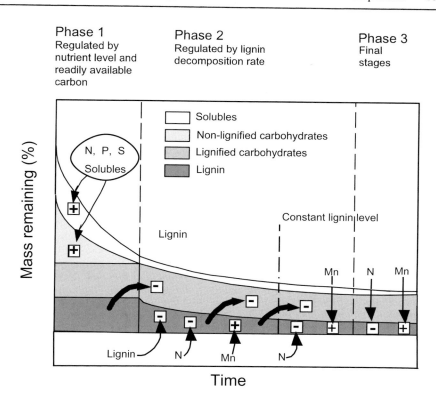

Fig. 6.1. Model for chemical changes and rate-regulating factors during decomposition, modified from Berg and Matzner (1997). The decomposition of water-soluble substances and unshielded cellulose/hemicellulose is stimulated by high levels of the major nutrients (early stage - phase 1). When all unshielded holocellulose is decomposed, only lignin-encrusted holocellulose and lignin remain. In this late stage (phase 2), the degradation of lignin rules the litter decomposition rate. N hampers the degradation of lignin and higher N concentrations suppress the decomposition whereas Mn appears to have a stimulating effect on the degradation of lignin. Finally, in the humus-near stage (phase 3), the lignin level is about constant, often at a level of 50-55%, the litter decomposition rate is close to zero and the accumulated mass loss also reaches its limit value.

6.2.2 Initial decomposition rates for newly shed litter - early decomposition stage in plant litter

Background

To determine rate-regulating factors within a given plot, studies are normally designed so that foliar litter types with different nutrient and lignin composition are compared in terms of mass loss over brief periods of no more than a year. When

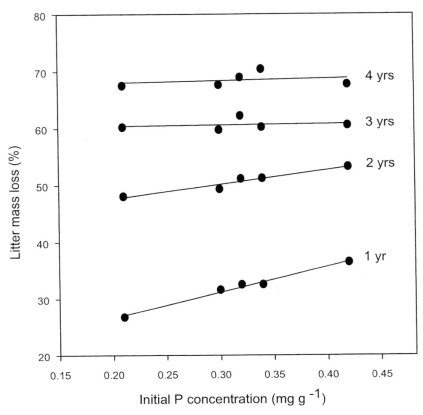

Fig. 6.2. Relationships between initial concentrations of P and increasing mass loss of Scots pine needles. A set of five collections of Scots pine needle litter was used from N-fertilized plots. The slope decreases as the rate stimulating effect of P decreases and that of another factor takes over (B. Berg unpubl.)

evaluating data the best linear relationship (highest R^2) should show the most limiting factor. However, not all foliar litter should be compared in this way, since nutrients in litter have different status and availability, and some newly shed litter, such as that from Norway spruce, may be in a late stage of decomposition even before it is shed.

How can initial rates be described?

The majority of studies on litter decomposition found in the literature present results from the early decay stages, where normally positive relationships are observed between litter concentrations of N, P or S, and the mass-loss rate or CO_2 release from litter. The concentration of water-soluble substances has also sometimes been related to initial rates.

Table 6.1. Overview of a provisional division of foliar litter types into groups of different properties, organized into the three-stage model (Fig. 6.1). The names of the groups are provisional but refer to those litter types characteristic for the properties

| Group | Stage of decomposition | | |
	Early stage	Late stage	Humus-near stage
"Pine"	Low initial leaching (<2%). Decomposition of solubles within the litter structure. Initial decomposition rates may be related to initial concentrations of N, P, S. Initial decomposition rates may be related to climate	Slow continuous increase in lignin concentration. Relationship between lignin concentration and mass loss rate.	Limit values normally above 80%.
"Deciduous"	High initial leaching (>5%). Rapid initial mass loss. Initial decomposition rates may be related to initial concentrations of N, P, S. Initial decomposition rates may be related to climate.	Maximum lignin level normally reached quickly. No relationship between lignin concentration and mass loss rate.	Limit values normally below 80%.
"Spruce"	Low initial leaching (<2%). Slow initial mass loss. Relationship between mass loss rate and nutrient concentration unclear. Influence of climate is probably small. Early stage is possibly missing.	Slow continuous increase in lignin concentration. Relationship between lignin concentration and mass loss rate not clear.	Limit values 70-90%.

There are different ways of expressing the decomposition rate in the early stage and a definition is useful (see Glossary and section 2.2). For litter types that leach their water-soluble compounds only to a small extent or not at all, decomposition for a given period (e.g. a month or a year) means microbial decomposition (C mineralization). For litters that leach C-compounds initially, for example several deciduous foliar litter types, mass-loss rates are the result of a combination of microbial decomposition and leaching (see Glossary).

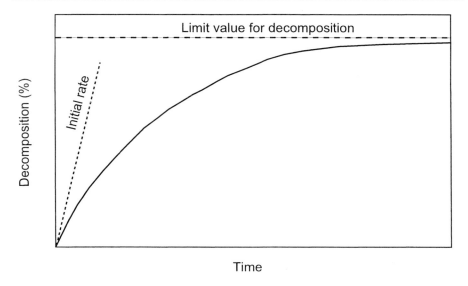

Fig. 6.3. Asymptotic model for describing litter mass loss during decomposition. Using extrapolation, the model can be used to estimate a quasi-steady state or limit value for accumulated mass loss

A further way to determine the initial rate of decomposition is to use the mass-loss data from a whole study (Fig. 6.3), ranging from the first sampling of litter-bags to the very last one, perhaps after several years. Equation (6.1) is a function that estimates the limit value for decomposition and also the initial rate as the derivative of the graph at time zero.

$$L = m(1-e^{-kt/m}) \qquad (6.1)$$

In this equation, L is the accumulated mass loss (in %) and t is time in days. The parameter m represents the asymptote at which the accumulated mass loss will remain constant, and the parameter k is the decomposition rate at the beginning of decay, sometimes indicated as k_{init} (the derivative of the function at $t=0$). To use this function, a set of litter mass-loss data over time is required. Often 10-12 samplings for a single litter type extending beyond 50% mass loss are required to give the sufficient data for both the limit value and the initial rate to be determined.

The loss of organic matter from newly shed litter through soaking the litter in water may reach rather high values, although this initial leaching can be limited in some litter types. For example, Scots pine needle litter may lose just a few percent from leaching (Bogatyrev et al. 1983), while some deciduous litters may leach considerably more, in part because they contain a higher concentration of water soluble substances (Table 4.1). The loss of soluble materials is not all due to leaching. In a study of sugar maple leaves on Black Hawk Island (see Appendix III), McClaugherty (1983) determined that most of the soluble carbohydrates and soluble phenols lost during the first year were either consumed by microorganisms or converted to insoluble compounds. After 1 year, only 21% of the initial soluble carbohydrates remained in the leaves, 15% had been leached and 64% had been respired or converted. For soluble phenols, the figures had a similar pattern: 29% leached, 4% remaining, and 67% consumed or converted. Freeze-thaw cycles increase the leachable amount both of C compounds and of nutrients.

Initial leaching is possibly more influenced by the litter type (coniferous *vs.* deciduous), freeze-thaw cycles, and concentration of solubles than by litter N levels. Fog (1988) suggested that litter N levels influenced the magnitude of the leaching (cf. Sect. 11.3.2). However, this may be more related to leaching in late stages than during initial decay. For a set of coniferous litter, Berg and Matzner (1997) presented data showing a negative relationship between initial levels of N and initial concentrations of water-soluble substances.

Initial chemical composition and different indices related to initial decomposition rates

Different approaches have been undertaken to identify a chemical index for the initial decay rate. One is simply the concentration of a given nutrient, such as N or P. A discussion about which of the main nutrients is rate-regulating in the early stage is not always meaningful, as other factors can be more dominant. We can see (Table 6.2) that for decomposing Scots pine needle litter, the main nutrients P and S show significant correlations to first-year mass loss, correlations that are emphasized when more nutrient-rich green needles were included in the measurements. That the relationships to N are weaker does not mean that N is without effect. The different R-values between N, P and S probably reflect differences in availability of the respective elements to the decomposer microorganisms. The lower R-value for the initial relationship to N could reflect the fact that part of N is stored in forms that are unavailable to the microorganisms that first invade the litter. In contrast to these three nutrients are the highly mobile ones, such as K, that are very leachable. Different initial concentrations of these more mobile nutrients appear not to influence the litter decomposition rate or their concentration in later stages of decomposition.

A positive relationship between a nutrient and initial decomposition rate is a crude measure of its rate-regulating function. Part of the N is tied to the lignin fraction (Flaig et al. 1959). About 1/3 of the total N is initially complexed to the lignin fraction in Scots pine (Berg and Theander 1984). Such N is then not readily available to the microorganisms that start the decomposition process. Aber et al. (1984) measured the amount of N associated with solubles, holocellulose and lignin fractions of six foliar litter species (Table 6.3). They found that between 26 and 38% of the total N was associated with the lignin fraction. This means that while a value for total N may be used as an index for available N it will overestimate the amount of available N. As a consequence this index may not be very reliable between species with different levels of available N. According to the literature, P and S do not appear to be bound like N, and as a result may be more available to leaching or microbial uptake (Stevenson 1982).

Table 6.2. Correlation coefficients (R) for linear relationships between first-year mass loss and initial concentrations of some main nutrients, water-soluble substances, and lignin as well as the lignin-to-N ratio. Scots pine and Norway spruce needles from N-fertilized trees were used for a within-species comparison and a set of different litter types for a comparison over species. Significance levels are given under the R values

	Scots pine[a]	Norway spruce[b]	Norway spruce[c]	Multiple species[d]
N	0.446	0.305	0.045	0.643
	ns	ns	ns	<0.01
P	0.904	0.556	0.063	0.797
	<0.001	ns	ns	<0.001
S	0.780	nd	nd	0.508
	<0.01			<0.05
K	0.899	0.511	0.126	0.649
	<0.001	ns	ns	<0.01
Ca	0.148	-0.693	0.032	0.161
	ns	<0.05	ns	ns
Mg	0.520	0.326	0.195	0.750
	ns	ns	ns	<0.001
Mn	nd	-0.226	0.570	nd
		ns	<0.05	
Wsol	0.217	0.888	0.265	0.792
	ns	<0.01	ns	<0.001
Lignin	-0.145	-0.663	0.122	-0.118
	ns	<0.1	ns	ns
/N	-0.650	-0.593	0.055	-0.773
	<0.05	ns	ns	<0.001
n	11	9	14	18

n Number of samples, *nd* not determined, *ns* non-significant, *Wsol* water soluble.
[a]Experimental Scots pine needle litter mainly originating from fertilized plots and with increased nutrient levels incubated at site Jädraås Data from Berg and Staaf (1982).
[b]Experimental Norway spruce needle litter originating from fertilized plots and with increased nutrient levels, incubated at the same plot. Data from Berg and Tamm (1991).
[c]Norway spruce needle litter incubated at 14 sites along Sweden with AET ranging from 371 to 545 mm. In that case no climatic influence could be traced. Data from Berg et al. (2000).
[d]Experimental Scots pine litter (above) as well as brown and green leaf litter from Scots pine, lodgepole pine, silver birch, and grey alder. Data from Berg and Ekbohm (1991).

The ash content of litter can vary between litter types and over time. Ash contents in sugar maple leaf litter were initially 11.3% of dry matter, increasing to 19.5% after 1 year of decay and 26.6% after 10 years of decay (C. McClaugherty unpubl.). If not considered, high ash contents could affect the calculation of the percentage of N and other substances relative to that of less ash-rich litter types. The N contents should thus be related to the litter organic matter, rather than to the whole litter, something that is done inconsistently.

The C-to-N ratio, which conceptually expresses N concentration in the organic matter, also gives a good relationship to mass loss for these early stage. This concept (C-to-N) is an index that originally was developed to be a rule-of-thumb for digestibility of fodder (e.g. fresh hay) but is today in use for soils as a means of predicting N dynamics (see Berg and Ekbohm 1983). For most species of newly shed litter, a low C-to-N ratio usually suggests an initially high decomposition rate.

Table 6.3. Initial N associated with extractable, acid hydrolyzable, and lignin fractions of six foliage litter species collected at Black Hawk Island, Wisconsin, USA. (Aber et al. 1984)

Litter	Extractable	Acid hydrolyzable	Lignin
		[mg g^{-1} OM]	
Bigtooth aspen leaves	4.0	1.5	2.8
Canadian hemlock needles	4.5	1.5	2.3
Red oak leaves	4.1	1.7	2.4
Sugar maple leaves	5.5	0.5	2.3
White oak leaves	4.7	0.7	3.0
White pine needles	2.2	0.0	2.2

OM organic matter

Another index is the lignin-to-N-ratio (Melillo et al. 1982). This ratio was based on the hypothesis that N and lignin had different effects on the decomposition rate, where N is a rate-stimulating and lignin a rate-retarding factor. The ratio is generally a good predictor of mass loss during initial stages of decay, particularly for litters from which it was derived (cf Table 6.2). For late stages (see below; Chap. 2) this index is of little value, since N for the late stages has a rate-retarding effect. The value of this index as a predictor of decay rate thus decreases as the decay process develops. The correlation between the lignin-to-N ratio and first-year mass loss is significant while the correlation between either lignin or N on their own and first-year mass loss is not.

The indices of initial chemical composition, which act as important rate-regulating factors, vary among litter types (Table 6.2). For Scots pine needle litter P and S were highly significant whereas N was not. When combining several litter types, levels of N, P, S and water-soluble substances all had significant predictive capacity. Potassium and Mg are neglected here as rate-limiting components, since their concentrations drop heavily immediately after incubation, and there has been no indication that they act as limiting nutrients. For Scots pine litter the lignin-to-N ratio was significant, although neither N nor lignin concentrations taken alone were. For a combination of different litter species incubated at the same site, N but not lignin was significant, while the ratio was highly significant, and better than N concentration taken alone.

For experimental Norway spruce needle litter the relationships between first-year mass loss and concentrations of lignin and water solubles were significant (Table 6.2; Berg and Tamm 1991), while P was not significant. There was no relationship between N concentration and initial rate of decay, and the lignin-to-N ratio was negative. For Norway spruce litter incubated in a transect (Table 6.2), no relationship was seen between climate and initial decay, and when comparing substrate-quality factors, only Mn was seen to be related.

For pine needles, and for different species combined (Table 6.2), concentrations of most of the main nutrients are positively related to litter mass loss. In addition, the concentrations of the main nutrients are normally positively correlated to each other. For example, N, P and S (Tables 4.2, 4.3 and 4.4) are all parts of the basic components of both plant cells and microorganisms, such as nucleic acids and proteins. However, only total concentrations of N, P and S are normally measured, without regard to availability. This makes any observed relationship between these nutrients more tenuous because the nutrients do not have the same availability.

At a Scots pine site, measurements of first-year mass losses were followed for several years for the most common litter type (Scots pine needle litter), with the annual variation among N, P and S given in Table 4.3. The most limiting element for microbial growth and activity can vary among years. Thus, a raised concentration of N may give a response in one year and not in another. Further, the ratios of N, P, S and other nutrients vary with time within the same stand and litter (Table 4.3). Cotrufo et al. (1998) found that rates did not increase when the initial concentration of N in litter was proportionately higher than concentrations of P and S, which were at more normal levels. We cannot exclude the possibility that the proportions of the main three nutrients are a critical factor that may vary among sites and litter types. Furthermore, N, P and S are not the only nutrients that influence the initial decomposition rate. Ca concentration has also been positively related to initial rates (Van Cleve 1974).

Although N appears to be one of several dominant rate-regulating factors in newly shed litter, the concept "newly shed litter" is neither simple nor clear. Some litter types show differing behaviors, and spruce needles, for example, deviate from the three-phase model (Fig 6.1). A possible explanation for the different decomposition pattern of spruce needles is that they are a more heterogeneous material that, in addition, is in a late stage of decomposition when shed. Because they remain attached to the twig for extended periods of time after they senesce, significant decomposition can occur while they are still attached to the tree.

Comments on a deviating foliar litter type: spruce

Two extensive studies on decomposition of newly shed litter of Norway spruce indicate a different behavior as compared to those litter types fitting the three-phase model. The decomposition rates of newly shed spruce litter are related to concentrations of water-soluble substances and lignin, rather than to the concentrations of the major nutrients (Table 6.2, Berg and Tamm 1991).

When Berg and Tamm (1991) compared the decomposition rates of needles of different chemical composition, there was no statistically significant relationship between initial mass-loss rate and concentrations of nutrients. It appeared though, that the initial concentrations of water solubles had a significant influence on mass-loss rates up to about 2 years of decomposition (Fig. 6.4), whereas the initial concentration of lignin had a less pronounced effect (although still statistically significant) on first-year mass loss. There was thus no nutrient-regulated early

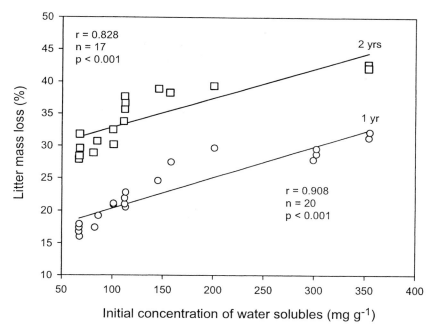

Fig. 6.4. Litter mass loss of Norway spruce needle litter as dependent on the initial concentration of water-soluble substances. Accumulated mass loss after 1 year, (○), and after 2 years (□) (Berg and Tamm 1991)

decomposition phase like that described above. This may be explained by the fact that the spruce needles stay on the branches after senescence for several months (H. Staaf, pers. comm.). During this period, leaching of nutrients could take place, and the decomposition of the needles could start, resulting in increased lignin concentrations and changed concentrations of other nutrients. The implication of this is that the early phase of decomposition may already have passed when the needles are shed.

In another study on Norway spruce needles throughout Sweden, carried out at 14 sites in a climate transect (cf. Chap. 7), there was no relationship between first-year mass loss and climate or climate-related indices for levels of AET ranging from 371 to 545 mm (Berg et al. 2000). When the data for the sites were combined and compared to substrate-quality factors, the concentrations of the main nutrients (N, P, S) did not give significant relationships, but the concentration of Mn correlated positively with first-year mass loss ($R^2=0.325$; $n=14$; $p<0.05$; Fig. 6.5). The important role of Mn in lignin degradation was described in Chapter 3.

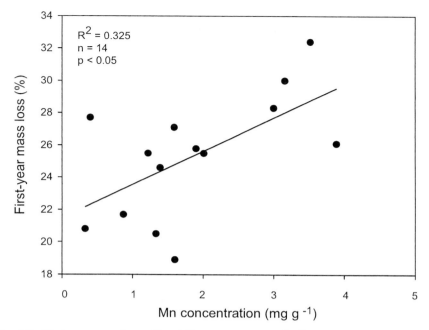

Fig. 6.5. First-year mass loss of local Norway spruce needle litter *vs.* initial concentration of Mn. The needle litter was incubated at 14 sites in a transect across Sweden. (Berg et al. 2000)

6.2.3 Decomposition in the late stage - lignin-regulated phase

The late decomposition stage

In an attempt to organize decomposition patterns, Berg and Staaf (1980a) distinguished a late phase in which the decomposition was regulated by lignin decomposition. They noted that when the effect of the main nutrients ceased, the rate could be related negatively to the increasing lignin levels. For Scots pine needle litter they estimated that the shift in phases took place at a mass loss of between 26 and 36% (Fig. 6.1). In a separate study on Scots pine needle litter Couteaux et al. (1998) determined the change in phases to be at about 25% mass loss.

Effect of lignin and N on litter decomposition rates

During decomposition, there is an increase in the concentration of the normally resistant compound lignin and its recombination products (Fig. 5.2). Traditionally, this has been explained by the fact that the lignin-degrading microorganisms normally grow very slowly, and that lignin as a chemical compound is normally resistant to decomposition, while the unshielded cellulose and hemicelluloses in litter are decomposed considerably faster. This resistance to degradation appears to be valid only under certain circumstances, however, and is partly regulated through the litter N level (Chap. 2) and the physiology of the lignin-degrading organisms present. Most studies on litter decomposition are on foliar litters where the levels of N have been high enough to influence the microbial degradation of lignin, thus creating an image of lignin as being more recalcitrant (Table 2.3).

When the decomposition has reached a certain magnitude the (foliar) litter contains material that is rich in lignin and its recombination products. At this stage the remaining cellulose and hemicelluloses are enclosed and protected by lignin and humins (cf. Chaps. 2 and 5). In late decomposition stages the degradation rates of cellulose and the different hemicelluloses are similar to that of lignin, whereas they are higher during earlier stages (Chap. 2). Thus, through the effect of N in the late decomposition stages, the degradation of lignin regulates the decomposition of the whole litter (Berg et al. 1987; Berg and Ekbohm 1991).

It is a well-known phenomenon that as litter decomposes, the N level also increases (Fig. 5.4). The rate of increase observed in N concentration is normally in proportion to the initial concentration (Chap. 5), meaning that the higher the initial N level, the greater the increase in N concentration with increasing mass loss.

The rate-retarding effect of lignin should be dependent on the N level of the litter, N exerting the suppressing effect on lignin degradation. We do not know whether the rate-retarding effect of increasing lignin levels should be ascribed to the simultaneously increasing N concentration in the litter. The same lignin level showed a stronger retardation at warmer and wetter sites than at colder and drier ones (Fig 2.8). The concentration of N increases relatively faster at warmer and wetter sites (Fig. 5.12), which means that higher N concentrations do occur in litter at sites where the proposed effect of lignin is stronger. The suppressing effect of N on the degradation of lignin, as well as on the decomposition of whole litter, has been observed in a variety of studies, and is based on both organic-chemical observations (Nömmik and Vahtras 1982) and on a microbiological-physiological level (Eriksson et al. 1990). The generally retarding effect of N should not be assumed to apply more generally across different ecosystems, but it has been observed and confirmed in the cases described here (cf Chap. 11).

Mass-loss rates of lignin as compared to initial litter N levels

It has been possible to distinguish differences in lignin degradation rates and relate them to litter N concentrations. These estimates of mass-loss rates of lignin were based on the measured values for sulfuric-acid lignin (Fig. 6.6A). It was observed earlier that sulfuric-acid lignin in decomposing litter was degraded at very different rates, notably in green N-rich and brown N-poor Scots pine needle litter. The lignin mass-loss rate was lowest for the N-rich litter and highest for the N-poor one (Berg et al. 1982b). Berg and Ekbohm (1991) also fitted a linear model including the N concentration of the litters and found a clear negative relationship between litter N concentration and lignin mass-loss rate (Fig. 6.6A,B).

We have focused on the suppression of lignin-degradation rates by N. There are other nutrients, though, that may influence the lignin-degrading ability of the mixed soil microflora. Such nutrients are likely to include Mn and Ca, although information exists only for pure cultures of lignin-degrading fungi.

The lignin fraction

Lignin in decomposing litter is a complex concept. Several methods to determine lignin were originally intended for fresh wood (for the paper pulp industry; e.g. Klason or sulfuric acid lignin). The applicability of such methods for other more heterogeneous substrates, such as foliar litter, both fresh and under decomposition, is not self-evident. The commonly used gravimetric determinations used in several methods may include components other than lignin, such as ash (containing a.o. Ca, Mg and silicates) and also recombination products derived from organic components. Still, gravimetric measurements are widely accepted, provided that they have been made correctly. However, unexpected effects have been observed during decomposition, and net increases in the lignin fraction have been reported (Berg and McClaugherty 1989). If for example, humic acids are synthesized and recorded as Klason lignin, the measured process of lignin mass loss should be regarded as a net process. Norden and Berg (1990) did not find any new peaks in the aromatic resonance region when applying high resolution ^{13}C NMR to needle litter samples in decomposition stages from 0 to 70% accumulated mass loss. Thus, within a certain range of litter mass loss, there does not appear to be any extensive synthesis of entirely new products. In addition, there was a clearly significant linear relationship between the lignin concentrations estimated using ^{13}C NMR and sulfuric-acid lignin.

The authors compared the lignin analysis according to Effland (1977) with that of Bethge et al. (1971) for fresh and decomposing litter and found no difference in concentrations (unpubl. research). Different data on lignin concentrations from different studies, using these methods, may thus be compared at least from a quantitative point of view. A forage analysis protocol introduced by Van Soest (1963) has also been widely used for lignin analyses. A study by Ryan et al. (1990) compared the Van Soest with the Klason technique as applied to deciduous litters.

Fig. 6.6. A Linear relationships for decomposition of sulfuric-acid lignin *versus* time in seven litter types of different initial N concentrations incubated in a 130-year-old Scots pine forest (Berg and Ekbohm 1991). **B** Illustration of the negative relationship between accumulated lignin mass loss changing with time and litter N concentrations. $R^2_{adj} = 0.677$, n=82. Data originates from seven litter types incubated experimentally in a Scots pine stand (Berg and Ekbohm 1991)

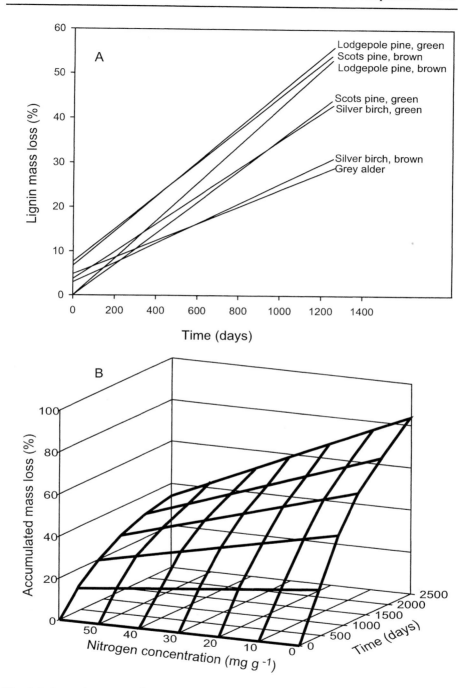

Fig. 6.6. Legend on preceeding page

They found that the two approaches gave somewhat different result, but that they could be interconverted (e.g. Aber et al. 1990). It is important to note that the terminology as regards "lignin" fractions and their contents sometimes is unclear (see Glossary), and that measurements of the lignin fraction using different methods may not give results that are directly comparable.

The biological regulation and the chemical mechanisms

We reviewed the biological and chemical aspects of lignin degradation in Chapters 2 and 3. As discussed above, elevated N levels may suppress the degradation of lignin and, at raised concentrations, the effect may be proportional to the N concentration. These two effects, biological and chemical, have so far not been separated at the level of litter decomposition, or the level of lignin degradation in litter. However, Berg and Matzner (1997) discussed effects of N additions to humus that suggested that both a biological and a chemical effect could be hampering the decomposition. In Table 2.3 we compared N levels in needle litter with N levels in pure laboratory cultures that were suppressing lignin degradation. In the relatively N-poor brown Scots pine needle litter there was between 40- and 100-fold higher N concentrations than was needed to suppress lignin degradation in a laboratory culture. Of course, much of the N in litter would be relatively unavailable to the fungi as compared to the N in the culture media. However, with initial concentrations of N in litters ranging up to nearly 3%, the N levels in litter are up to 800 times as high as concentrations that have a suppressing effect on lignin degradation in pure cultures of fungi.

The effect of added N on respiration rate from humus was seen within hours (Berg and Matzner 1997), indicating a microbial mechanism. We may expect that a repression of fungal ligninase synthesis occurs also in litter, as a result of the relatively high levels of N present. The chemical reaction between N and the decomposing lignin is slow at the low pH values (around 4) in boreal needle litter. Still, in a laboratory experiment the reaction proceeded at a rate of 14-19 μg N day^{-1} g^{-1} litter (Axelsson and Berg 1988). The reaction rate was limited by N availability, and using Scots pine needle litter as a substrate, Axelsson and Berg (1988) found that the reaction rate increased with increasing N concentrations. Thus, with the long-term decomposition in the field we can speculate that over a period of years the reaction between N and lignin will become prominent.

How should we regard the retardation of litter decomposition caused by lignin?

Often the "effect of lignin" on decomposition rate has been illustrated by showing the correspondence between increasing lignin concentrations and decreasing mass-loss rates. A higher concentration of lignin would thus reflect a higher percentage of a compound resistant to decomposition, with a rate that is dependent on the litter N concentration, and the kind of organisms that have invaded the litter. We can distinguish some cases that represent the extreme possibilities, but our discussion is based on incubation of litter on the forest floor, not in laboratory systems.

If N-sensitive white-rot fungi invade the litter and dominate, we should find that the N concentrations, increasing with accumulated mass loss, are hindering the degradation of lignin to an increasing extent. Thus the overall decomposition of the litter will be increasingly hindered. In this case raised N levels could result in a lowered litter decomposition rate. The white-rot fungi as a group have the ability to degrade lignin fast when N levels are low, so any effect of N should be clearly distinguishable (Chap 3.). An effect like this may be expected in a nutrient-poor system in which N-sensitive fungi dominate.

Another possibility is that white-rot fungi that are not sensitive to raised N levels invade the litter. The lignin degradation of such a population would not be hindered by raised N concentrations, and lignin would thus not be a barrier to litter degradation at higher N concentrations. In such cases, either there would be no correlation between raised lignin levels and decomposition rate, or perhaps more likely, there would be no increase in lignin levels.

Brown-rots would not degrade lignin completely, and after the disappearance of the unshielded holocellulose, the raised lignin concentrations would hinder litter decomposition. This would apply to both N-sensitive and N non-sensitive species.

Considering the decomposition of foliar litter, the most likely scenario when litter is incubated on the ground is invasion by a mix of fungal species. For example, Osono and Takeda (2001) found over 100 different fungal taxa on the decaying leaves of Japanese beech. Thus, we would expect that both sensitive and non-sensitive white-rots and brown-rots would participate in the degradation. Such a mix of species would result in a moderate suppression of lignin degradation at low N levels, while higher N levels would have a stronger effect. Differences between systems would be reflected in the slope of the decomposition rates for lignin and for litter (Fig. 6.6). Thus, a high initial litter N level would have a stronger rate-retarding effect on litter decomposition than a litter with lower N incubated under the same conditions. We also speculate that a system richer in N would have relatively more fungi not sensitive to N, while really nutrient-poor systems would have a relatively high frequency of N-sensitive fungi, thus allowing a stronger retardation of litter decomposition in the latter type of system (Eriksson et al. 1990; Hatakka 2001).

Based on the cases reported in the literature, suppression is normal for foliar litter. In fact there may be just one case reported for which the lignin degradation has not been hampered. In a paper on lignin decomposition in beech leaf litter, Rutigliano et al. (1996) reported that lignin concentrations decreased right from the start of the study. Later the concentrations started to increase slowly from a level much lower than the initial one. The system was very rich in nutrients (the humus had ca. 3.7% N in the organic matter; Berg et al. 1996b), and it is possible that the system was dominated by white-rot fungi that were largely unaffected by raised N levels.

Some different lignin-related decomposition patterns among litter types

We previously mentioned three groups of patterns that appear to be characteristic for the early and late stages of foliar litter decomposition. We gave them provisional names relating to groups of species, namely "pine", "deciduous", and "spruce" and discussed them in that order (Table 6.1). Here we intend to relate the lignin dynamics to these groups.

In the cases reported, lignin is resistant to degradation and an increasing lignin concentration suppresses the decomposition rate of the litter in most foliar litter types. The relationship shown above, namely a decreasing rate for one type of litter incubated at its own forest stand, has been observed by several scientists. One basic method to investigate the possible effect of a chemical component on decomposition rate, is to incubate the litter over a series of years, and regard the litter that changes with decomposition as a new substrate at the beginning of each incubation year. The mass loss for an individual year is thus associated with litter chemical composition at the start of that year. In this approach, the lignin concentration at the start of each one-year period is regressed against the mass loss over that one-year period to obtain a slope for each site describing the effect of lignin. The effect of an increasing lignin level in foliar litter and its decomposition rate is thus easily illustrated.

Pine needle litter. The mass-loss rate of needle litter is retarded in proportion to lignin concentrations (Fig. 6.7A), an observation that has been made repeatedly in Scots pine ecosystems as well as in ecosystems of other pine species (Berg and Lundmark 1987; McClaugherty and Berg 1987). In the case of two pine species, Scots pine and lodgepole pine, with similar N levels and growing in systems of the same soil richness (Fig. 6.7A), we note that the relative lignin-mediated effects are similar, with a slow and even increase in lignin concentration during the decomposition process.

Deciduous type. Whereas the pine litter gave negative linear relationships for mass loss *vs.* lignin for the late stage with rather slow stepwise changes, such linear relationships are not seen for the deciduous litter types that were studied. There is a clear effect of lignin that can be illustrated by the initial (often first year) decomposition and those of the following years. After a rather quick initial decomposition and leaching phase, the lignin concentration quickly reaches a maximum value that causes a lower degradation rate, creating a graph with two clusters (Fig 6.7B) of points rather than one giving a continuous change, as is the case for pine needle type litter. The late stage, for deciduous litter represented by a cluster of points, is thus clearly different from that of e.g. pine litter and no relationship to lignin concentration is evident.

Spruce type. Norway spruce needle litter appears to deviate from the above two types, in which the rate-retarding effect ascribed to lignin has been noted to start after lignin concentrations have increased from an initial value, and after all unshielded holocellulose is degraded. Berg and Tamm (1991) compared the effect of lignin concentration on litter mass-loss rate for individual incubation years and found significant relationships only in the first year, and none in years 2, 3 and 4

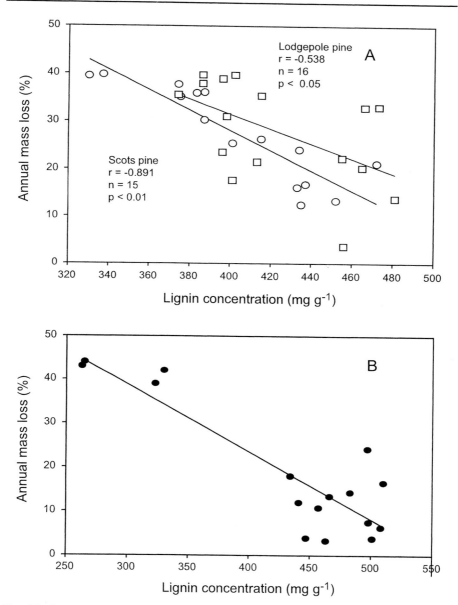

Fig. 6.7. Relationship between lignin concentration at the beginning of a 1-year decay period and the mass loss occurring during the following 1-year period. **A** Scots pine needle litter (○) and lodgepole pine litter (□), both litter species in the late decomposition stage. **B** Leaf litter of silver birch and grey alder including both early and late stage. The litter in the late stage forms a cluster with lignin levels above 430 mg g^{-1}

(Table 6.4). When investigating the late phase, and comparing annual mass loss and lignin concentrations, they found that the concentration of lignin had a rate-regulating effect on litter decomposition when present in a concentration range from 350 to 450 mg g^{-1}, causing a decrease in annual mass loss from 24 to 10% (Fig. 6.8, Table 6.4). Above a lignin concentration of 500 mg g^{-1}, the deviation in annual mass-loss values increased considerably with no pattern evident. Berg et al. (1987) noted a similar phenomenon, though less pronounced, for Scots pine needle litter in a very late stage of decomposition.

Table 6.4. Correlation coefficients (R) for the linear relationship between annual mass loss of Norway spruce needle litter and initial concentrations of lignin at the start of each one-year period. Analyses are based either on incubation year or grouping of litter vs. lignin concentration (cf. Fig. 6.8) (Berg and Tamm 1991)

Variable	R	n	p
Incubation year			
1st yr	-0.894	10	<0.001
2nd yr	-0.482	11	ns
3rd yr	-0.234	11	ns
4th yr	-0.376	8	ns
Lignin conc. range			
<450 mg g^{-1}	-0.873	16	<0.001
<475 mg g^{-1}	-0.774	22	<0.001

ns Not significant

Considering that there was no lignin effect above 450 mg g^{-1} and that the lignin concentration did not increase much above that, we can conclude that at high concentrations of lignin, near its maximum value, other rate-regulating factors begin to emerge (Berg et al. 1984; Chap. 5). Such a conclusion would also imply that the effect of N on the lignin-degrading microflora might be overshadowed. Alternatively, the reactions between N and lignin degradation products that build a chemical barrier may have ceased, possibly due to saturation by N. We may speculate that in these late stages the influence of some other component may dominate the lignin degradation.

In a transect study with locally collected Norway spruce needles, only a part showed an increasing lignin concentration followed by a decreasing annual mass loss, thus following the pattern described for pine (Berg et al. 2000). Lignin concentrations correlated negatively with litter decay rates for seven out of 14 sites. For the seven other sites no such lignin effect was seen. Thus two groups of sites could be distinguished (see also transect No 4 in Sec. 7.5).

Berg et al. (2000) combined the data within the two groups of sites (Table 6.5). Group I sites had significant relationships between lignin concentration and annual litter mass loss. Group II sites showed no significant relationships between lignin concentration and mass loss. The litter of Group I (n=55) gave a highly significant, negative relationship between annual mass loss and concentrations of lignin (Fig. 6.9A, Table 6.5). For Group II (n=33) only the relationship between mass loss and litter Mn concentration was significant (R^2=0.277; p<0.01; Fig. 6.9B).

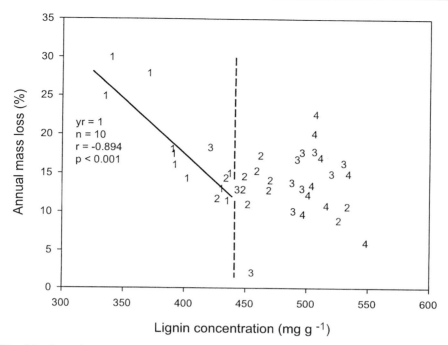

Fig. 6.8. Annual mass loss of Norway spruce needle litter as compared to lignin concentration in litter at the start of each one year period. Data originates from an experiment in N fertilized plots (Berg and Tamm 1991) and from two control plots at which N-fertilized litter was incubated. The *number* gives the approximate incubation year. The vertical line gives an approximate limit for the effect of lignin

For this group, there was no significant relationship between annual mass loss and concentrations of lignin, water solubles, N, P, K, Mg or Ca (Table 6.5).

The effect of Mn becomes even more pronounced when combining all available data for annual mass loss of Norway spruce needle litter, the first year excluded. The best linear relationship was obtained for annual mass loss *vs* Mn concentration ($R^2=0.356$, $n=95$; Table 6.6). For the entire data set, neither N nor lignin had a significant effect.

We have already discussed (Chap. 2) the effect of N (Eriksson et al. 1990) and Mn (Perez and Jeffries 1992; Hatakka 2001) on lignin degradation. Lignin decay and therefore litter decomposition rates may be limited if one or more of the essential elements required for microbial degradation of lignin are lacking. Likewise, high concentrations of an element such as N could suppress microbial degradation of lignin. Such nutrient interactions may be complex, but the composition and activity of the microbial community, including the lignin-degrading fungi, depends greatly on concentrations of nutrient elements. It appears possible that the differing concentrations of Mn in litter could be dependent on site (soil) properties (Berg et al. 1995a).

Table 6.5. Linear regressions for combined data of decomposing Norway spruce needle litter. Data from a transect study with 14 sites (cf. Chap. 7). At seven sites, negative, significant relationships were seen between lignin concentration and annual mass loss. The data from these sites is combined into group I (n=55). Group II is formed from data from sites at which no significant relationships were seen between lignin concentration and annual mass loss (n=33). (Berg et al. 2000)

	Group I (sign. relationships)			Group II (non-sign. relationships)		
	R	R^2	p	R	R^2	p
Lignin	-0.775	0.600	<0.001			ns
Wsol	0.673	0.453	<0.001			ns
N	-0.608	0.37	<0.001			ns
P	-0.498	0.24	<0.01			ns
K	0.330	0.109	<0.05			ns
Mg	0.554	0.307	<0.001			ns
Mn	0.316	0.100	<0.1	0.526	0.277	<0.01
Ca	0.281	0.079	<0.1			ns

ns Non-significant, *Wsol* water-soluble substances.

Table 6.6. Linear relationships between all available data for annual mass loss of Norway spruce needle litter and concentrations of chemical components in the litter. The data covered all litter of a Norway spruce transect (Berg et al. 2000) and that from an experimental site and included mass losses during the second through the fifth years. Concentrations of N and lignin did not have significant relationships (B. Berg unpubl.). n=95

Variable	R	R^2	p<
Mn	0.597	0.356	0.001
Mg	0.487	0.237	0.001
Wsol	0.454	0.206	0.001
K	0.367	0.135	0.001
Ca	0.254	0.065	0.05
P	0.230	0.053	0.05
All components	0.712	0.507	0.001

Wsol Water solubles.

6.2.4 Litter at a humus-near or limit-value stage

General comments

Limit values for decomposition may be calculated for many litter species, provided that decomposition has been followed far enough through the decay process (Chap. 2). After an inventory of existing decomposition studies, Berg (2000b) and Berg et al. (1997) published a total of 128 limit values of which 106 originated from forest sites that were natural and not disturbed. When calculating values for different litter types, the limit values were found to differ (Howard and Howard 1974; Berg and Ekbohm 1991). Berg et al. (1996b) related limit values negatively to initial litter N concentrations and positively to initial Mn concentrations. They presented the hypothesis that the amount of litter remaining at the limit value was regulated by lignin remains that had been recalcitrant due to fixation of low-molecular weight N compounds that was in turn enhanced by the raised litter N levels. The relationships were empirical and suggestions for causal relationships included factors that influence lignin degradation and modify lignin structure. These were attempts to build a theory about the nature of the recalcitrant remains. Currently, we can state that the empirical relationships have been found to be more general and the recalcitrance of the remaining litter has been validated, but still there is no clear theory. In a search for possible factors regulating the limit value, relationships have been found with litter concentrations of N, Mn, Ca and lignin, all of which have a potential causality. Reviews containing increasing numbers of limit values have been published (Berg et al. 1996b, 1997; Berg 2000b) and the main patterns observed have remained.

General relationships

Higher N levels in the litter give a lower decomposition. When the existing 106 limit values for foliar litter decomposing in natural systems were regressed against concentrations of nutrients and lignin, it was seen that N concentration gave a highly significant, negative relationship (Fig. 6.10, Table 6.7). The possible reasons for the relationships between limit values and N were discussed in Chapter 2. The relationship using all available data from natural systems was negative ($R^2 = 0323$; n=106; p<0.001) meaning that the higher the initial concentration of N the lower the limit value, with a smaller proportion of the litter being decomposed. Behind this observation there may be a causal relationship that is valid both for litter in late decomposition stages and for humus. The discussion that was applied onto 'litter in late stages' (cf. Chap. 2) may also be used in this case. After further inventories, 128 limit values have been estimated (Berg 1998a,b) and they still have a highly significant relationship to initial N levels.

The fact that in this large data set the relationship to N concentration was significant indicates a general effect of N over a good number of species in deciduous and coniferous ecosystems in boreal and temperate forests. In addition, initial litter N concentrations ranged from 0.3 to 3.0%. Although significant, the R^2 value was still low (0.324) when including all 128 limit values. This probably depends on the fact that in this broadly based data set, there are several factors potentially influencing the limit value, which increased the variability of the data. Since data were collected for different forest ecosystems, with litter being incubated on soils with different properties and in different climate zones this is not strange. In the large data set, several species were represented and Berg (2000b) calculated the average limit value for the eight dominant species and compared this to their average N concentration (Fig. 11.3, Table 11.5). Analysis of this subset of data showed a clearer relationship between initial N levels and limit values. The litters in the complete data set were often significantly different as regards both N level and limit value (see also Chap. 11). As different forest humus systems have different levels of nutrients this may be expected. The large difference between two systems is illustrated in Table 11.3, namely a Scots pine system with a humus N level of 11.8 mg g^{-1} and a silver fir system with an N level of 38.2 mg g^{-1} and a

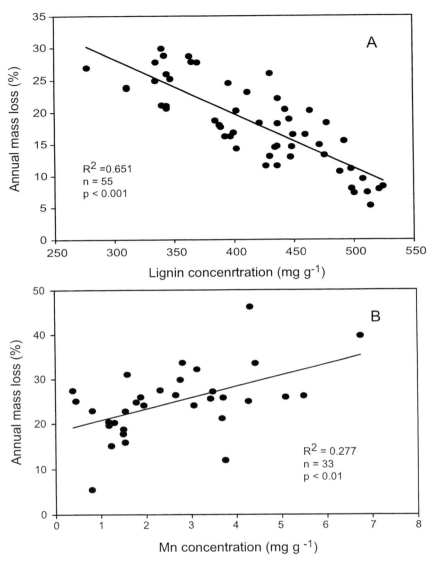

Fig. 6.9. Annual mass loss plotted *vs.* substrate-quality factors at the start of each year for local Norway spruce needle litter in late decomposition stages incubated in 14 Norway spruce forests across Sweden. Lignin and Mn concentrations at the start of each 1-yr period were related to the litter mass loss in the following 1-yr period. **A** Annual mass loss *vs.* lignin concentration at those seven sites where negative linear relationships between lignin concentration and annual mass loss were seen. **B** Annual mass loss *vs.* Mn concentration at those seven sites where no relationship between annual mass loss and lignin concentration was seen (cf Tables 6.5 and 7.9). From Berg et al. (2000)

Table 6.7. Correlations between limit values, and initial concentrations in litter of N, P, S, K, Mg, Mn, Ca and lignin respectively. With data from all sites pooled the extent of the chemical analyses determined the amount of data in the statistical analysis of each nutrient (different *n* values). (Berg and Johansson 1998)

Nutrient / litter group	R	R^2	n	p<
Natural systems – all data				
N	-0.568	0.323	106	0.001
Mn	0.519	0.269	83	0.001
All deciduous litter				
N	-0.438	0.192	30	0.05
Mn	0.618	0.382	13	0.05
Ca	0.675	0.456	18	0.01
All coniferous litter				
N	-0.660	0.436	86	0.001
Mn	0.513	0.263	74	0.001
Scots pine needle litter alone				
N	-0.683	0.466	42	0.001
Mn	0.485	0.235	35	0.01
Norway spruce needle litter				
Lignin	-0.742	0.551	11	0.01
Ca	0.636	0.404	11	0.05

generally higher level of the other nutrients. With such differences, the soil microorganism populations would have adapted to the different levels.

Higher Mn values are related to higher limit values. When the relationship between limit values and initial Mn concentrations of the litters was investigated, Berg et al. (1996b) showed a clearly significant, positive relationship (R^2=0.529; *n*=17; *p*<0.001; cf. Chap. 2) for natural undisturbed systems. With 83 limit values, litter Mn concentrations gave a highly significant, positive relationship (R^2=0.269, Table 6.7). For selected groups of litter such as deciduous (*n*=13) and coniferous (*n*=74), positive relationships were seen between Mn and limit values with R^2=0.382 and 0.263, respectively. In addition, for Scots pine needle litter alone a significant relationship was seen (R^2=0.235), whereas limit values for spruce needles did not have any significant relationship to Mn (Table 6.7).

Higher lignin values are related to lower limit values. Using all available limit values, Berg et al. (1997) noted a negative relationship to initial lignin concentrations in litter. Such a relationship may be expected, as lignin appears to be an important component in the nucleus of the recalcitrant part. When all litter types were investigated, the relationship was weak, with an R^2 value of 0.044 for 112 measurements, whereas for single ones, e.g. spruce litter, the relationship was stronger (Table 6.7) with R^2=0.551. The relationship between lignin concentration and limit values is not necessarily a simple one and we cannot exclude the possibility that this relationship is based on lignin as a mediator.

Heavy metals. A very N rich system could be expected to have a higher percentage of lignin-degrading organisms that are not sensitive to N (cf. Chap. 3). This could mean that the limit values are ruled to a lesser extent (or not at all) by the concentration of N, while other factors such as the levels of heavy metals may

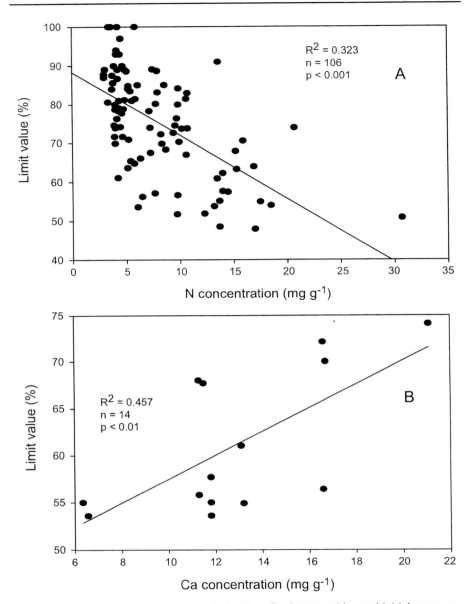

Fig. 6.10. Linear relationship between limit values for decomposition and initial concentrations of N and Ca in foliar litter. **A** Available data from natural forest systems plotted *vs.* litter N concentration. (Berg 2000b). **B** Available data from Norway spruce forests plotted *vs.* litter Ca concentration. (Berg 2000a)

be important. Heavy metals have been measured and relationships have been noted in data originating from two very contrasting sites (Berg et al., in press). No clear global relationships were apparent from these studies, however negative relationships between limit values and litter Cd and Zn concentrations were seen at a moderate level of significance ($p<0.1$).

Nutrient and heavy metal concentrations as indicators. The fact that significant relationships exist between limit values and initial concentrations of N and Mn suggests a rate-regulating mechanism. However, until a causal mechanism is demonstrated, the effects of initial concentrations of these elements should be viewed as empirical indices only.

These indices should be regarded in different ways. For a nutrient such as N, concentrations increase linearly with accumulated mass loss, and mainly in proportion to the initial concentration (see Chap. 5; Berg et al. 1999b), thus the use of initial N concentration should not cause any problem when used as an index. Similar reasoning may be applied for some heavy metals. For heavy metals such as Cu, Pb and Fe the concentrations in later stages reflect the initial ones as their concentrations increase (cf. Chap. 5). However, for some nutrients and heavy metals, with Mn as an example, mobility is pH dependent. With these elements an increase does not normally take place during decomposition and it remains to be determined how we should interpret the relationship between limit values and their concentrations. A specific effect as regards Mn *vs.* lignin degradation has been observed in laboratory studies. However, these studies have not led to a clear understanding of such effects in more complicated soil systems.

Groups of litter have different empirical relationships

The observed relationships between N and Mn concentrations and limit values were also seen for selected groups of litters (Berg et al. 1996b, 1997). Berg and Johansson (1998) made a first division into coniferous and deciduous litter, however, such a division into litter types disregards the effects of site quality (cf. below).

The coniferous litters as a group gave a highly significant relationship between limit values and litter N concentrations ($R^2=0.436$, $n=86$, $p<0.001$; Table 6.7) as did Mn ($R^2=0.263$, $n=74$). Berg (1998a) also found enough studies using Scots pine to allow a special investigation of the factors regulating the limit value for that specific litter species, and found a highly significant, negative relationship between N concentrations and limit values ($R^2=0.466$, $n=42$, $p<0.001$). Needle litter from fertilized trees with raised levels of N and other nutrients, followed the same general pattern, with the best relationship being to N. Manganese, Ca and lignin gave significant relationships as well.

For Norway spruce needles from undisturbed systems, a highly significant, negative relationship was found between limit values and the concentrations of lignin ($R^2=0.551$, $n=11$, $p<0.01$, Table 6.7) and Ca in litter. However, there was no relationship to either N or Mn concentrations. When investigating spruce needle litter from N-fertilized plots Berg (2000a) found relationships between both litter Ca and Mn concentrations and limit values.

For deciduous litter as a separate group, the limit values were positively related to litter concentrations of Mn (R^2=0.382, n=13, $p<0.05$) and Ca (R^2=0.456, n=18, $p<0.01$; Table 6.7).

Do limit values indicate a stable fraction? Can we describe the properties of the recalcitrant remains? Do limit values indicate a more or less complete halt in the litter decomposition process? Although limit values for litter mass loss have been estimated for a variety of litters by using asymptotic functions, we cannot conclude that such limit values indicate that the remaining organic matter is completely recalcitrant to degradation by biological agents (see below). Instead, the residual organic matter could very well consist of a moderately stabilized fraction that decomposes very slowly or a fraction that just does not decompose in a given environment, whereas a changed environment or disturbance could allow further degradation. However, this would not mean that the discovery of an apparent final mass-loss value should be considered trivial, especially if the limit value could be related to climate and litter properties, such as lignin concentration, nutrient status or other environmental factors.

Couteaux et al. (1998) applied both a three-factorial model (Chap. 10) and a limit value function to direct CO_2 measurements of decomposing Scots pine needle litter close to the limit value, and to the humus formed in the same stands. They measured k-values for decomposition of a stable fraction in the magnitude of 0.0001 to 0.00001% per day. These k-values correspond to a decomposition rate of about 1% per 30 and 300 years, respectively. That study included an analysis of stable, meta-stable, and labile components (Table 11.6), where the stable fraction comprised ca. 80% of the material and may be considered as rate limiting.

The estimated limit values may thus illustrate a fraction that is highly stabilized and thus decomposes at a very low rate. Even if this should be the case, the limit value concept is no less useful, especially if we can connect this recalcitrance in litter to its chemical properties, for example to the initial concentrations of lignin or some nutrient, or possibly to climate (cf. Chap. 11).

6.3 Does chemical composition influence leaching of compounds from humus?

In the literature new observations indicate the existence of a disintegration or decomposition mechanism for humus that appears to be initiated by high acid or high N-loads in the soil system. Very high N loads, for example in N deposition, appear to promote a disintegration of humus, probably as a consequence of heavily increased microbial activity. This theory was forwarded by Fog in 1988. He suggested that a higher concentration of N in litter/humus would result in an increased production of soluble organic matter (DOM or DOC). His ideas were based on the functioning of the group of lignin-degrading organisms that are called "soft-rots" (Chap. 3). Worrall and Wang (1991) briefly reviewed the literature and added their own observations to show that at least some soft-rot fungi need high levels of N in their surroundings in order to carry out decomposition.

Therefore, in an N-rich environment they can, to a certain extent, replace white-rot organisms. Soft-rot degradation of lignin yields remains of incompletely degraded lignin that react with organic N compounds, a reaction that leads to water-soluble products. Fog's (1988) conclusion was that high N concentrations increase the formation of water-soluble compounds, resistant to degradation, but decrease the amount of humus that is formed, as might occur in a mor layer. Ulrich (1981) has described a similar process and called it a "disintegration of humus". David et al. (1989) reported higher concentrations of soluble organic matter with increasing acidity.

Guggenberger (1994) concluded that the mobilization of DOC is not ruled exclusively by a low pH. On the contrary he makes the reasonable conclusion that high inflows of total N suppress the complete lignin degradation carried out by white-rot organisms but increases the general microbial activity. He supports the conclusion by Fog (1988) that the more N-tolerant soft-rot fungi produce partial degradation products that are more water soluble, especially the N-containing compounds. He also proposes that a generally higher microbial activity will give a greater production of microbial metabolites.

We may compare this to the isolation of N-tolerant white-rot fungi (Chap. 3) from N-rich surroundings and conclude that the phenomenon described by Guggenberger (1994) and Fog (1988) could be due simply to a wider spectrum of fungal species. The common property among these fungi could be a tolerance of high N levels. Connected to the above observations is a comparison of amounts of humus in mineral soil under Douglas-fir and red alder. Cole et al. (1995) found that in the more N-rich alder stand, greater amounts of C compounds were leached into the mineral soil from the humus layer. While it is difficult to base wider conclusions on this study alone, the observation that an N supply in a naturally richer environment is part of a mechanism for formation of DOC that later precipitates in the mineral soil, does fit Fog's (1988) hypothesis.

7 Climatic environment

7.1 Introduction

Climate has a dominant effect on litter decomposition rates on a regional scale, whereas litter quality dominates on a local level (Meentemeyer 1984). Thus, at a given site and climate, one should expect differences in mass-loss rates of litters to be related primarily to their chemical and physical properties. Many studies have demonstrated such relationships (Fogel and Cromack 1977; Aber and Melillo 1982; Upadhyay and Singh 1985; McClaugherty et al. 1985; Dyer 1986). As the decay of litter progresses through time, the constituents that regulate the rate of mass loss can change. Berg and Staaf (1980a) presented a schematic model of these litter decay stages, later modified by Berg and Matzner (1997; Chap. 6). Thus early-stage decomposition is primarily controlled by concentrations of limiting nutrients, especially N and P, whereas lignin decomposition exerts the dominant control in the latter stages. The delivery of heat and moisture to the litter will exert a control over the rate at which the decay phases postulated by Berg and Staaf (1980a) can proceed. Thus in one climatic regime the early, nutrient-controlled, phase could persist while in other regimes this phase could be quickly passed (Dyer et al. 1990).

Analyses of decay dynamics have been conducted using widely different litter types, at sites in different climatic regimes and in different forest types. Thus, when considering the regulation of decomposition it is difficult to separate the influence of climate from the influence of litter quality. With the increasing emphasis on understanding the impact of climate changes on the broad scale patterns of biological processes, we need to expand our consideration of the decomposition processes from substrate and site specific studies to a broader regional context. At this larger geographic scale our attention must turn to climate.

In this chapter we have focused on litter decomposition in stands with monocultures and use results from five main climatic transects with either Norway spruce, or solely Scots pine or a mixture of Scots pine and other pine species. We suggest that the results are sufficiently contrasting to illustrate that different patterns should be expected among species in climatic transects.

7.2 Microbial response to temperature and moisture

The soil microbial community encompasses several hundred species in the soil of a particular stand (Torsvik et al. 1990; Bakken 1997). The microbial community has a rich adaptability to both different moisture and temperature regimes (Chap. 3), but both moisture and temperature can be limiting. Unless there is enough moisture, often above about 10% water-holding capacity, water may be so limiting that raising temperatures would not result in higher microbial activity. Likewise, in an energy-limited system, such as at low temperatures, higher moisture would not necessarily result in higher activity.

The microbial response to temperature should be regarded as the sum of responses from the entire microbial community. Those bacteria and fungi having their temperature optima at say, 15 °C are less active at 10 °C and nearly inactive closer to freezing point. However, at zero and even below, psychrophilic organisms carry out a clear heterotrophic activity. These organisms belong to completely different species from those active at higher temperatures. In systems in cold climates the microorganisms would thus be adapted to the prevailing climatic conditions.

A microbial response to variability in climate would thus be dependent on the availability of both nutrients and carbon sources (see review by Panikov 1999). A lack of an essential nutrient or available carbon source relative to the needs of the microbial population would thus result in a lack of response to a variation in climate. In situations where substrate quality poses a severe limitation on decomposition, changes in climate can have relatively little effect on decomposition rates.

7.3 Early-stage decomposition of Scots pine needle litter

7.3.1 Decomposition at one site

Within a single site there is a clear variation in litter decomposition rates among years that may be due to a variation in weather. When Scots pine needle litter was incubated at its own site, the first-year mass loss as determined over 19 measurements ranged from 21.1 to 33.8% (Fig. 7.1), giving a range factor of 1.6. There was no difference in annual mass loss between litter samples incubated in the spring and in late autumn just after litter fall. Average values for both sets were close to the general average value of 27.8% year^{-1} mass loss.

However, there are differences in decomposition rates between periods of the year determined by temperature and rainfall patterns and intensity. A model was constructed that related soil water and temperature quite well to litter mass loss (Jansson and Berg 1985; Fig. 7.2) with R^2 values ranging between 0.85 and 0.99, indicating that the annual variation in climate may dominate the annual variation in mass-loss. The model calculated soil climate on a daily basis. When soil water

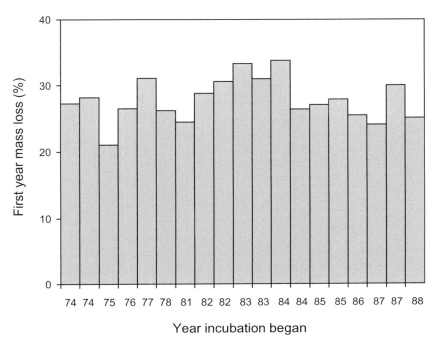

Fig. 7.1. First-year mass loss from Scots pine needle litter incubated in an 120-150-year-old Scots pine forest at the former site of the Swedish Coniferous Forest Project (SWECON) at Jädraås, Sweden. The first incubation was made in 1974 and the latest in 1988. In some years incubations were made in both spring and autumn (B. Berg unpubl.; B. Andersson unpubl.)

and soil temperature were combined into a single factor, this had clearly superior predictive power than either of the single factors alone (Table 7.1).The model predicted the soil climate over a period of 6 years that included a large variety of weather patterns. Thus, soil moisture and temperature also exhibited large variability. Two summers were characterized as warm with extended drought periods, whereas the other summers were moist. The variations in soil temperature were much more pronounced between different winters than between summers. Of the winters, three had soil temperatures well below zero degrees, which also caused high water tension in the soil. The other winters were both moister and warmer, mainly because of thicker snow packs, which prevented the soil from becoming completely frozen. Under these conditions the soil water was always unfrozen, which means that decomposition took place under the snow cover. In fact, for one particular 1-year period, the main part of the decomposition took place during the winter. As indicated earlier, the initial chemical composition was variable (Table 4.3), but the model, based on climate, explained decomposition quite well.

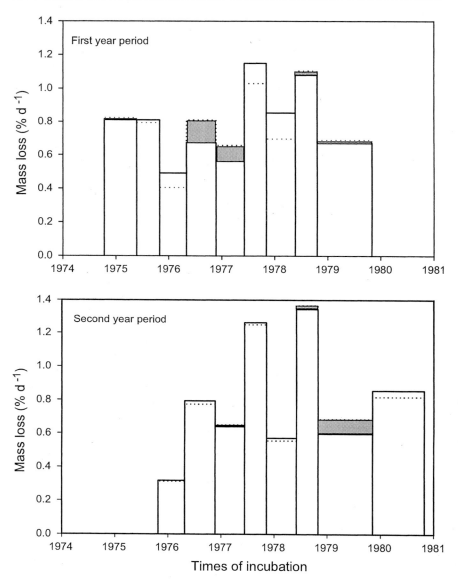

Fig. 7.2. Observed decomposition rates (——) for periods longer than 145 days for the first and second year of decomposition compared with predicted litter mass loss (•••••) based on soil climate output from a soil-climate model. *Shaded areas* show underestimated mass loss. (Jansson and Berg 1985). The study was undertaken in a nutrient-poor Scots pine monocultural forest, ca.120-130 years old

Table 7.1. Coefficients of determination (R^2) obtained from correlations between observed decomposition rates and different soil climate estimates as independent variable. Scots pine needle litter was used and incubated annually at its own site. (Jansson and Berg 1985)

Independent variable	Year of incubation		
	First	Second	Both
Actual evapotranspiration (AET)	0.41	0.74	0.55
Soil temperature	0.37	0.77	0.52
Soil water tension	0.78	0.97	0.81
Soil water content	0.68	0.96	0.77
Soil temp. and water tension	0.90	0.98	0.89
Soil temp. and water content	0.85	0.99	0.87

7.3.2 Decomposition over transects

We will review several major studies on decomposition in different climatic transects, in Northern Europe. Five of the studies investigated needle litters and one investigated root litter:

— Transect 1 with Scots pine, located mainly on glacial till in which local Scots pine needle litter was incubated once or twice. Twenty-two stands/sites were located between 66°08'N, near the Arctic Circle and 55°39'N, at about the latitude of Copenhagen in southern Scandinavia (Tables 7.2 and 7.4).

— Transect 2 with Scots pine on sediment soil, in which unified Scots pine needle litter was incubated annually for periods lasting from 6 to 19 years. The transect had 13 sites between northernmost Finland (69°45') and central Holland (52°02'N). This transect had highly standardized sites with nutrient poor Scots pine stands on sandy sediments and thus on flat ground. At each of these sites, however, a special set of experimental litter was incubated (Table 7.7).

— Transect 3 was composed of pine forests located mainly on sediment soils in which unified Scots pine needle litter was incubated. Transect 1 was included and sites with stands of Austrian pine, Corsican pine, stone pine, maritime pine and Monterey pine. This transect included 39 sites ranging across Europe from northernmost Finland (69°45'N) to southernmost Spain at (38°07'N) and southernmost Italy (39°24'N) (Tables 7.5 and 7.6, Figures 7.3, 7.4, and 7.5).

— Transect 4 with Norway spruce stands located on glacial till in which local litter was incubated once. Fourteen sites were used, located between 66°22'N, close to the Arctic Circle in Scandinavia, and 56°26'N in southernmost Sweden (Tables 7.8, 7.9 and 7.10).

— Transect 5 with Scots pine. This transect had sites along a west to east gradient of continentality. Sites were located between 52° and 53°N and from Berlin in the west (12°25'E) to the Russian/White Russian border in the east (32°37'E).

— A transect with root litter in a northeast to southwest transect that included 25 stands with Scots pine and lodgepole pine and 12 stands with Norway spruce.

The sites ranged from near the Arctic Circle in Scandinavia to Berlin ($52°28'N$) (see Sect. 7.6, Table 7.12).

The decomposition data from the Transects 1-5 and that for root litter were related to both climate and substrate-quality data, the former calculated according to Meentemeyer (1978). The climate indices are listed in Table 7.3 with their acronyms along with acronyms for substrate quality.

Table 7.2. Calculated slope coefficients for the relationship between annual mass loss and increasing lignin concentration in litter at the start of each 1-year period (cf. Figs. 2.8, 7.6). Data are taken from a transect with local Scots pine needle litter incubated at sites ranging from the Arctic Circle in Scandinavia to Lüneburger Heide ca. 100 km south of Hamburg, Germany. (Johansson et al. 1995)

Site	slope	SE	R^2	R	n	p<
2 Harads	-0.0231	0.0144	0.076	-0.276	33	ns
3:1 Manjärv	-0.0216	0.0421	0.036	-0.189	7	ns
3:2 Manjärv	-0.060	0.0597	0.173	-0.416	9	ns
3:3 Manjärv	-0.132	0.0209	0.278	-0.527	8	0.05
4:23 Norrliden	-0.0815	0.0217	0.453	-0.673	19	0.01
6:51 Jädraås	-0.0734	0.0240	0.227	-0.476	34	0.01
17:2 Kappsjön	-0.1751	0.0473	0.774	-0.880	6	0.05
18:2 Anundberget	-0.1874	0.0551	0.794	-0.891	5	0.05
103:1 Tomta	-0.045	0.0593	0.055	-0.235	12	ns
102:1 Kungs-Husby[a]	-0.107	0.0353	0.568	-0.754	9	0.05
105:1 Remningstorp[a]	-0.166	0.043	0.65	-0.806	10	0.01
101:1 Grensholm[a]	-0.148	0.0518	0.577	-0.760	8	0.05
107 Sänksjön[a]	-0.166	0.043	0.65	-0.806	8	0.01
8 Nennesmo	-0.230	0.0516	0.665	-0.815	12	0.01
10:1 Mästocka	-0.228	0.0533	0.901	-0.949	4	ns
13 Ehrhorn	-0.250	0.0334	0.846	-0.920	12	0.001

ns Not significant, *SE* standard error.
[a]Set contains some data from N-enriched litter.

Scots pine monocultures

Local litter. When investigating the data of Transect 1 ranging over Scandinavia, Johansson et al. (1995) determined the effect of climate and litter-quality variables on mass-loss rates. They used long-term average climatic values when relating first-year mass loss to climate variation (Table 7.4), and found that average annual temperature (AVGT) gave the best fit with a value for R^2_{adj} of 0.518. Annual actual evapotranspiration (AET) gave almost as good a fit with an R^2_{adj} of 0.505. Potential evapotranspiration and average temperature in July were also significantly related to mass loss, but annual precipitation, water surplus, and   water

Table 7.3. Climate and substrate quality variables towards which litter mass loss was regressed in studies of litter decomposition in transects (transects nos. 1-4 and transect with root litter). The climate variables were calculated according to Meentemeyer (1978) and Thornthwaite and Mather (1957). For convenience these acronyms are used in this chapter

Activity Acronym	Description of variable
LOSS	Annual mass loss of litter (%) either as first-year or as annual mass loss in later decomposition stages
Climate variables	
JULT	Average temperature for July (°C)
AVGT	Average annual temperature (°C)
PRECIP	Total annual precipitation (mm)
PET	Potential annual evapotranspiration (mm)
AET	Actual annual evapotranspiration (mm)
SUR	Soil moisture surplus (mm)
DEF	Soil moisture deficit (mm)
Substrate quality variables	
WSOL	Concentration of water solubles at the start of each year (mg g^{-1})
LIGN	Lignin concentration, initial or at the start of each year (mg g^{-1})
NITR	Nitrogen concentration, initial or at the start of each year (mg g^{-1})
PHOS	Phosphorus concentration, initial or at the start of each year (mg g^{-1})
POTA	Potassium concentration, initial or at the start of each year (mg g^{-1})
CALC	Calcium concentration, initial or at the start of each year (mg g^{-1})
MAGN	Magnesium concentration, initial or at the start of each year (mg g^{-1})
MESE	Manganese concentration, initial or at the start of each year (mg g^{-1})

deficit did not give any significant relationships. AET had previously been distinguished as a superior climate index at broad, continental scales (Meentemeyer 1978, 1984; Berg et al. 1993a,b). As the R^2_{adj} value for AET was close to that obtained using annual average temperature, Johansson et al. (1995) used it as a basis for further analysis.

On this scale, they found no relationships between first-year mass loss and initial concentrations of water solubles, N, P and lignin. None of these factors was significant, probably because the variation in climate across the 28 sites overwhelmed the control by substrate quality. Thus, for this litter type the first-year mass loss supports the traditional image.

Unified litter. In a more specific investigation, a set of sites with numerous observations was investigated. This set of 13 sites in Scandinavia and the northwestern part of continental Europe had mass-loss measurements over a period of between 6 and 19 years (transect 2). Of the single climate factors, AET gave a highly significant relationship ($R^2_{adj}=0.867$, $p<0.001$; Table 7.5). It is likely that the multiple years of observations gave an average calculated mass-loss rate more representative of the climatic norms used in this study as compared to the one-time incubation in transect 1. Part of this transect (10 sites) was placed in standardized Scots pine forests on level ground and nutrient-poor sediment soils. This part of the transect gave a fit to AET at the same level as that of all the 13 sites ($n=10$; $R^2_{adj}=0.866$; $p<0.001$).

Table 7.4. Linear relationships between first-year litter mass loss, and climate factors in a climatic transect from the Arctic Circle in Scandinavia (north-east) to the latitude of Copenhagen in the south-west (transect 1). Local Scots pine needle litter was incubated at 22 stands/sites. All climate and substrate quality variables listed in Table 7.3 were tested ($n=28$). (Johansson et al. 1995). R^2_{adj} means an R^2 value compensated for different degrees of freedom between comparisons

Climate factor	Slope (SE)	Intercept (SE)	R^2_{adj}	$p<$
AVGT	2.729 (0.498)	20.869 (5.812)	0.518	0.001
AET	0.134 (0.025)	-30.162 (5.890)	0.505	0.001
PET	0.143 (0.027)	-37.522 (5.955)	0.493	0.001
JULT	3.871 (1.787)	-28.6645 (7.856)	0.120	0.05

Substrate-quality factors alone did not give any significant relationships but the inclusion of NITR or WSOL as a substrate-quality index improved the relationship somewhat; for AET plus NITR an R^2_{adj} value of 0.885 was obtained (Table 7.5). The addition of other climatic factors added very little to a explain the observed variation. This part of the world's boreal forests are energy limited (Berg and Meentemeyer 2002) and the relationships may be improved by selected temperature functions.

Monocultural pine stands of different species

All sites. Using 39 pine sites (transect 3) Berg et al. (1993a,b) incubated needle litter samples in regions with AET ranging from ca. 330 to 950 mm and with a highly standardized site selection and design. The sites ranged over differing climates across western Europe from boreal northernmost Finland to southern Spain and the contrasting subtropical southern Georgia (USA). Unified litter was incubated two or three times a year at the different sites. First-year mass-loss ranged from about 10% at the northern-most boreal site to 56% at the subtropical one.

The best positive correlations were obtained for the relationship between first-year mass loss and AET ($R^2=0.509$) and the total annual precipitation (PRECIP, $R^2=0.323$) both with $p<0.001$, and average temperature (AVGT, $R^2=0.203$, $p<0.05$). Of the climatic variables, water deficit (DEF) also gave a small but significant correlation (Table 7.6).

First-year mass loss was plotted against the best single variable (AET) using all sites (Fig. 7.3). The progression in rates from the arctic to the subtropical sites is readily apparent. Some of the scatter can be attributed to the use of long-term climatic means rather than information about the weather during incubation. For example, the Georgia sites should have higher rates of mass loss in normal years than those that were observed, because incubation occurred in an extremely dry year. Some of the variation must also be caused by variations among local site conditions and litter quality. Although the litter originated from the same site (unified litter), there were some differences in chemical composition among years. The N concentration in the incubated litter ranged from 3.4 to 4.1, P from 0.19 to 0.21, S from 0.32 to 0.38, and Ca from 2.3 to 7.1 mg g^{-1} (Table 4.3).

Table 7.5. Coefficients of determination for linear correlations between first-year mass loss of unified Scots pine needle litter and selected climatic factors, as well as some sub-strate-quality factors. Sites, from transects 2 and 3, were grouped and investigated separately as well as in combinations of groups. All correlations presented here are significant at $p<0.001$. For acronyms see Table 7.3. (Berg et al. 1993a)

Independent variables	R^2	R^2_{adj}	Types of pine stands in transect
Scandinavian – northwestern European sites ($n=13$)			
AET	0.878	0.867	Scots only
AET, NITR	0.895	0.885	
Scandinavian- continental Europe north of the Alps and the Carpathians ($n=23$)			
AET	0.647	0.630	Scots only
AET, WSOL	0.748	0.736	
Scandinavian - northwestern European sites plus Atlantic sites ($n=22$)			
AET	0.916	0.912	Scots, Austrian, Monterey, and maritime
Mediterranean plus central European and North American sites ($n=17$)			
AET	0.753	0.736	Scots, stone, Monterey, and red
AET, WSOL	0.766	0.750	
AET, JULT	0.761	0.745	

Table 7.6. First-year mass loss of unified Scots pine needle litter in 39 pine forests (transect 3) as a function of some single climatic factors as well as multiple ones. A broad regional scale was used across Europe from an arctic site close to Barents Sea to southern Spain and southern Italy and included subtropical sites in southern Georgia, USA. Acronyms from Table 7.3. (Berg et al. 1993a)

Independent variables	R^2	R^2_{adj}	p	Comments
AET	0.509	0.496	<0.001	
PRECIP	0.323	0.304	<0.001	
AVGT	0.203	0.181	<0.01	
PET	0.187	0.165	<0.05	
DEF	0.097	0.073	<0.05	DEF gave a negative relation
AET, JULT	0.689	0.681	<0.001	JULT gave a negative relation
AET, JULT, AVGT	0.716	0.708	<0.001	JULT gave a negative relation

Atlantic climate sites. This climate, with rainy summers and moderate winters, encompasses all Scandinavian-northwestern European sites. The 13 highly standardized sites in the long-term Scandinavian transect (Transect 2), the sites of transect 3 with an Atlantic climate ($n=7$), and data from two eastern Finnish sites, had similar relationships between mass loss and AET (Fig. 7.4). All of these sites had low water deficit with the exception of one that was located very close to the coast. With these similar responses the 22 decomposition values were combined for an analysis *vs.* climate. Comparing AET and first-year mass loss gave a very good fit (R^2_{adj}=0.912; Fig. 7.4A). This relationship was not improved by the addition of other climatic factors or substrate quality.

Sites with warm, dry summer climate. The relationships obtained for the Atlantic climate sites were significantly different from those for Mediterranean and inland sites. Although AET had similar ranges for the two climate types, the pattern and temporal distribution of temperature and precipitation was of importance. Combining the mass-loss values for the sites characterized by dry and warm summers resulted in a set of sites encompassing those with a Mediterranean climate, sites in central Europe, and in the American Midwest. A linear regression of mass loss *vs.* AET gave a significant relationship (R^2_{adj}=0.736, n=17, p<0.001; Fig. 7.4B). The relationship was not improved by climatic factors indicating little contribution to the relationship by seasonality or by substrate-quality factors.

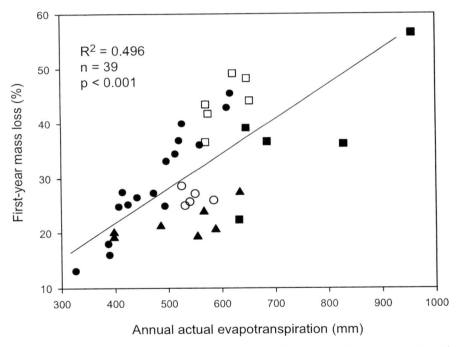

Fig. 7.3. Bivariate plot of average first-year litter mass loss *vs.* actual annual evapotranspiration (AET). (●) Scots pine sites in an intensively studied Fennoscandian-NW-continental transect ranging from northern Finland to central Holland; (□) pine sites close to the European west coast or sites relatively exposed to Atlantic climate; (▲) pine sites around the Mediterranean; (○) central European Scots pine sites (Poland) with characteristics of inland climate; (■) pine sites in the eastern inland of the United States. (n=39; Table 7.5) (Berg et al. 1993a)

Latitudinal transect. In a different approach, Breymeyer‘ and Laskowski (1999) investigated a latitudinal transect (Transect 5) with increasing degrees of continentality ranging from Berlin in the west (12°25′E) to the Russian/White

Russian border in the east (32°37′E). This transect was oriented along a gradient of increasing continentality. With increasing continentality the annual average temperature decreased, temperature amplitudes and precipitation increased, while in the same direction, the first-year mass loss decreased. Their experiment thus indicated that the distribution of the climate over the year is of importance for mass-loss rate, which was also seen in the comparison of Atlantic and inland climates (above).

Comments on regional comparisons. Berg et al. (1993a,b) showed that general broad-scale climatic control of mass loss rates in pine needle litter could be modeled. Their results also show that regions have differing responses that may be related to climate patterns. This means that the slopes and intercepts of the relationship can vary between different climates (Fig. 7.4). Effects of climate patterns may be direct or indirect. The forest floor environments are indeed different under pine forests of different regions, even though macro-climatic AET values are similar. Increasing continentality may result in changes in the composition of ground vegetation (Roo-Zielinska and Solon 1997, 1998), which may change the ground climate and other environmental conditions important for decomposition.

In this comparison, climatic variables that respond to seasonality and continentality were included, but none of these variables could help explain lower rates in the Mediterranean and inland sites. The mix of years and sites used suggests that this is not simply the result of experimental error (Berg et al. 1993a). Furthermore, the results using the Fennoscandian and Atlantic sites are very similar to those found by Meentemeyer and Berg (1986) using earlier data sets for Fennoscandia and weather records for the actual incubation period. Their regressions using AET *vs.* needle litter mass loss had intercepts and slope coefficients as well as R^2 values similar to those found here. The functional basis for these differences among climatic zones remains unclear.

7.3.3 Soil warming experiments

As part of the international Climate Change Experiment (CLIMEX), Verburg et al. (1999) incubated downy birch leaf litter in southern Norway in soils with temperature elevated by 3 to 5 °C. They found no effect of soil warming on litter mass remaining over a one-year period and concluded that the increased temperature was offset by decreased moisture.

In the Adirondack Mountains of northeastern New York, USA, McHale et al. (1998) incubated litterbags containing leaf litter of American beech and sugar maple. The bags were incubated for 2 years in plots maintained at 2.5, 5, and 7.5 °C above ambient soil temperatures during the snow-free season. Decomposition of beech leaf litter did respond to temperature, with the mass loss after 2 years significantly greater in the 7.5 °C plots than in the control plot (58 *vs.* 41% mass loss). Sugar maple leaves showed smaller differences between the same plots with 58 *vs.* 51% mass loss, respectively.

Rustad and Fernandez (1998) conducted a similar experiment in Maine, USA, with red spruce and red maple foliage litter, using a 5 °C warming. Red maple

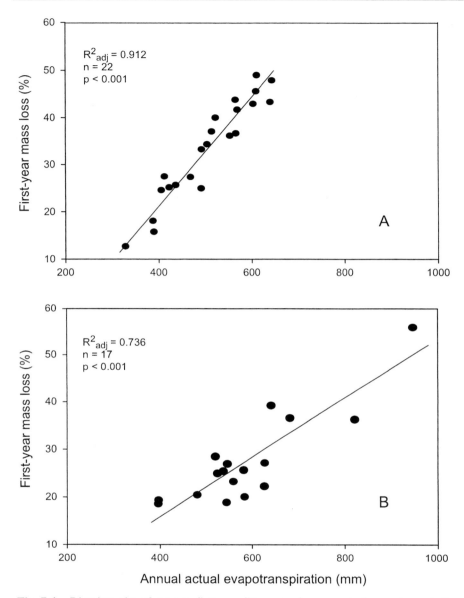

Fig. 7.4. Bivariate plot of average first-year litter mass loss *vs.* actual evapotranspiration. **A** Scots pine sites in a Fennoscandian-NW-continental transect and pine sites close to the European west coast, relatively exposed to Atlantic climate (*n*=22). **B** Mediterranean sites, central European ones and North American ones (*n*=17). A unified Scots pine needle litter was used. (Berg et al. 1993a), cf Table 7.5

leaves showed a slightly increased mass loss during the first 6 months of decay, but after 30 months, the elevated temperature had no effect. Red spruce needles behaved in the opposite way, with significant differences in mass loss showing up only after 30 months of incubation. Unfortunately, these studies took place during a particularly dry period and moisture may have been limiting in the control as well as in the heated plots.

Both these latter studies led their authors to suggest that temperatures will have the greatest effect on the early stages of decay, when labile materials are being degraded. In later stages, as litter quality becomes limiting, temperature seems to have less of an impact.

7.4 Effect of substrate quality on mass-loss rates in Scots pine transects

7.4.1 Early stage

At a given site, different litter materials decay at rates that are largely dictated by their chemical and physical properties (Berg and Staaf 1980a; Berg and Ekbohm 1991). These relationships may be unique to a site and its decomposer organisms. Therefore, predictions of decay rates for other sites cannot be made with confidence on the basis of the effect of substrate quality at a single site. The analysis of decay dynamics at a site must include the combined effects of both climate and litter-quality variables.

In part of transect 2, litters of different qualities were incubated at 11 sites (Table 7.7). This transect included sites in Finland, Germany, The Netherlands and Sweden. For each site the litter-quality variables (concentrations of N, P and water-soluble constituents) were regressed against annual mass loss (Berg et al. 1993a). Most of the regressions, even considering the low number of litter types at each site, were significant ($n=4$, $p<0.1$). Examination of the intercepts and slope coefficients for each regression equation at each site, suggested a consistent change in coefficients that was influenced by climate, similar to what was found by Dyer (1986).

Berg et al. (1993a) analyzed the set of intercepts and slope coefficients for the 11 sites with respect to each of the climatic variables. Their analysis revealed the degree to which the coefficients vary with climate. For concentrations of both N and P, the intercepts were strongly and positively related to annual potential evapotranspiration (PET), and the slope coefficients were related to precipitation at the site. Thus the intercepts appear to be driven mostly by annual temperature (temperature is a main substituent of PET), and the slopes of the relationship (mass loss *vs.* quality) by the gross water supply (precipitation).

Table 7.7 Chemical composition of four experimental Scots pine and lodgepole pine needle litters incubated at the sites in transect 2. (Berg et al. 1993a)

Litter type	Ws	Es	Lig	N	P	S	Mg	Ca	Mn	K
					Concentration [mg g^{-1}]					
Scots pine										
Brown, natural	164	113	231	4.8	0.33	0.55	0.49	4.42	0.79	1.07
Brown fertil	135	91	265	7.0	0.33	nd	0.37	2.50	0.70	1.02
Green natural	199	63	284	13.4	1.47	0.98	0.85	2.82	0.41	4.90
Lodgepole pine										
Brown natural	103	42	381	3.9	0.34	0.62	0.95	6.35	0.95	0.56

Ws Water soluble, *Es* ethanol soluble, *Lig* lignin, *fertil* fertilized, *nd* not determined.

These relationships were described empirically using derived linear equations based on data from a boreal to temperate climatic transect (Berg et al. 1993a).

$$\text{Mass loss}_{Phos} = (-29.3+0.111(PET))+(0.749+0.013(PRECIP))(PHOS) \quad (7.1)$$

Equation (7.1) shows the relationship between mass loss rates and PET and initial phosphorus concentration. The first term in parenthesis is actually an intercept determined by the site's PET (mm). The second term is the slope coefficient and is determined by the annual precipitation at the site (mm). The initial P concentration (PHOS) is the independent variable.

$$\text{Mass loss}_{Nitr} = (127.3+0.100(PET))+(-0.067+0.0022(PRECIP))(NITR) \quad (7.2)$$

Equation (7.2) is completely analogous to Eq. (7.1) and demonstrates the relationship between climatic variables, N concentrations (NITR), and annual mass-loss rates.

The expanded model for the influence of initial concentrations of N and P at any particular site may be written as nomograms (Fig. 7.5, Eqs. 7.1 and 7.2). This nomogram was constructed from Equation 7.1. Selected PET values are shown on the left vertical axis, annual precipitation on the horizontal axis and predicted mass-loss rates on the right vertical axis. The figure provides predicted loss rates for PET values between 400 and 600 mm, variable precipitation (from 200-650 mm) and initial P concentrations of 0.15(a), 0.30(b), 0.60(c), and 1.20(d) mg g^{-1}. Thus, PET determines the intercepts and precipitation (PRECIP) determines the slopes. Using this nomogram, the mass loss rate at a given site can be predicted on the basis of initial P concentrations.

These relationships (Fig. 7.5; Eqs. 7.1 and 7.2) also suggest that most of the regional variation in early-stage mass-loss rates in Scots pine forests across northern Europe is driven by temperature/heat constraints (Berg and Meentemeyer 2002). As precipitation increases, the differences in mass-loss rates for litter of differing P concentrations became larger. The sites used in this investigation all had an Atlantic climate (cf. above) and we could expect that the corresponding relationships for Mediterranean and continental sites would be different.

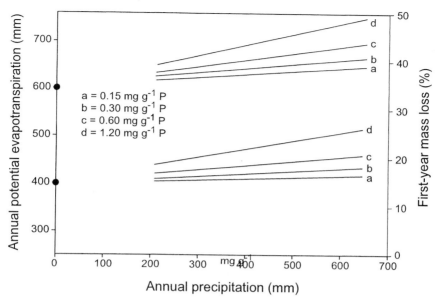

Fig. 7.5. Relationships among annual mass loss (%), potential evapotranspiration (PET), annual precipitation, and initial P concentration in litters. To investigate the effect of differing litter qualities under different climates, four chemically different litter types were incubated at 11 sites (part of transect 2, Table 7.7). Nomogram constructed from Eq. (7.1). Selected potential evapotranspiration (PET) values are shown on the *left vertical axis*, annual precipitation on the *horizontal axis* and predicted mass-loss rates on the *right vertical axis*. The figure provides predicted loss rates for PET values of 400 mm (*lower set of graphs*) and 600 mm (*upper set of graphs*) and four initial P concentrations of 0.15 mg g^{-1} (**a**), 0.30 mg g^{-1} (**b**), 0.60 mg g^{-1} (**c**), and 1.20 mg g^{-1} (**d**). (Berg et al. 1993a)

Figure 7.5 illustrates an alternative approach to comparing the roles of climate and litter quality in determining mass loss rates across a large geographical area. Even small changes in climate can produce greater changes in early-stage decay rates than large differences in litter quality. Thus, it is not surprising that in this type of system, quality variables are important at local scales but their influences are less significant when viewed over a larger scale. Nevertheless, the equations described here should permit predictions of the influence of litter quality across a broad area of northern European pine forests, especially those with Scots pine.

7.4.2 Late stage

Lignin concentrations increase during decomposition of litter (cf. Chap. 5), and raised lignin concentrations have been related to decreased litter decomposition rates (Fogel and Cromack 1977). The rate-dampening effect on litter mass-loss rates acting through lignin concentrations can be described with a negative linear relationship (Berg and Lundmark 1987), which for pine litter, may begin as early as 20% mass loss. In earlier work, Meentemeyer (1978) and Berg et al. (1993b) related mass-loss rates to lignin concentrations and demonstrated a variation with geographic location.

Johansson et al. (1995) calculated slopes for the relationship between lignin concentration and annual mass loss in a 2000-km-long climatic transect (part of transect 1) and found negative relationships. The steepest slopes were obtained for the southern sites that were warmer and wetter and, thus, had initially higher mass-loss rates than the more northern ones (Fig. 7.6). In fact, for three dry, nutrient-poor northern sites, the slopes became so shallow that the R^2 values became very low (Table 7.2). Thus, the slope for the southernmost site, no. 13 at Lüneburger Heide, Germany (Fig. 7.6) was -0.250% mg^{-1} g^{-1}, but a value of -0.023% mg^{-1} g^{-1} was determined close to the Arctic Circle in Sweden. The slopes for the sites in south and central Sweden were in-between (Fig. 7.6). The lengths of the lines define the intervals between highest and lowest lignin concentration used for the relationship.

The range in lignin-concentration values could influence the slope. However, Johansson et al. (1995), using two sets of data (for sites 13 and 6:51; Fig. 7.6) to isolate a range with the same lignin-concentration interval, made a comparison of the slopes and found that they remained the same. In a further step they used the slopes at each site and compared them to climate. They performed a second set of linear regressions and found that the best fit was that between slope and AET (Fig. 7.7), with an R^2 of 0.559. Other climatic variables gave significant correlation with slope as well, e.g. PET and annual average precipitation with values for R^2_{adj} of 0.413 and 0.405, respectively. They also combined all data in a multiple regression analysis with AET, lignin and their product as independent variables. This analysis showed a strong significance for all three terms in the model, and an adjusted coefficient of determination (R^2_{adj}) of 0.346 ($n=196$; $p<0.0001$), thus explaining 34.6% of the variation. This offers support for the conclusion that the relationship between litter mass-loss rate and lignin at a site is dependent on, or related to, the values of the climatic factors, especially AET. However, this relationship is empirical and the underlying causal factors are not known.

The causal mechanisms behind the relationship between lignin concentration and mass-loss rate could depend on the litter N concentration (Chap. 6). That the mass-loss rates were affected more strongly by increasing lignin concentrations in warmer and wetter climates, means that the degradation of lignin was more hampered at such stands. The litter N concentration increased more quickly in litter incubated in stands located at higher AET (Chap. 5), which may be a partial explanation. Furthermore, in the initial stages, the same litter types incubated at the same sites responded very differently towards climate (Fig. 7.4). We can reiterate that conditions that are initially rate stimulating may, in later stages, become rate inhibiting.

It appears that when litter has entered late decomposition stages, its decomposition rate is affected more by increasing lignin concentrations at sites with higher AET values, namely under warmer and wetter conditions. It also seems that the

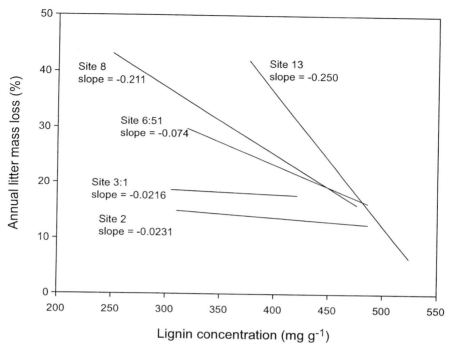

Fig. 7.6. Plot of annual litter mass loss *versus* litter lignin concentrations at the start of each year. The forest stands ranged between the Arctic Circle in Scandinavia and Lüneburger Heide ca. 100 km south of Hamburg, Germany (Table 7.2)

slopes of the lignin *vs.* mass-loss relationship tend to converge at a lignin concentration a bit higher than 500 mg g^{-1} litter, suggesting that the mass-loss rates approach a similar value at this level of lignin concentration, irrespective of climate. Thus at very late, humus-near stages, decomposition rates would not be driven principally by climate factors.

An experiment by Dalias et al. (2001) may confirm this observation. They investigated the effect of different temperatures during decomposition on the decomposability of the residual litter substrate. Using humus from five coniferous sites in a transect from 64°00' N in Sweden to 43°07' N at the Mediterranean, they incubated a ^{14}C-labeled straw material at 4,16 and 30 °C. They let the straw decompose to the same level of mass loss as measured through the released $^{14}CO_2$. Then the material was reincubated and the release of $^{14}CO_2$ showed that the highest mineralization rate took place in samples that had been originally incubated at 4 °C and the lowest in those originally incubated at 30°C (Fig. 7.8). Their interpretation was that when litter decomposed under higher temperatures, its residual compounds were more recalcitrant.

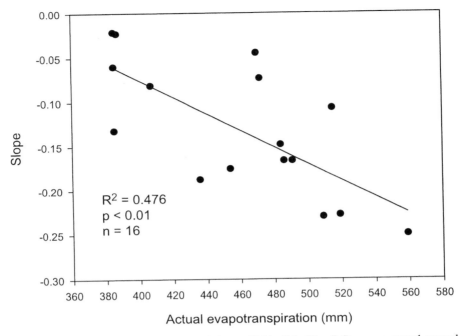

Fig. 7.7. Bivariate plot of slope coefficients (Table 7.2; Fig. 7.6) versus actual annual evapotranspiration (AET) for 16 sites in a climatic transect. The slope coefficients originate from the relationship between litter lignin concentration and annual litter mass loss. From Johansson et al. (1995)

7.5 Climate and decomposition of Norway spruce needle litter

7.5.1 Climate *versus* first-year mass loss

Over a climatic transect (Transect 4), the decomposition rate of Norway spruce needle litter was more closely related to substrate quality than to climate. Norway spruce needle litter is a substrate with properties very different from those of different types of pine needle litter (cf. Chaps. 4 and 6), and over the transect, this was reflected in a switch from climate control to one of substrate quality. For example, in a north-south transect along Scandinavia from the Arctic Circle (66°22′N) to the latitude of Copenhagen (56°26′N, climate indices did not show any significant relationship to first-year mass loss.

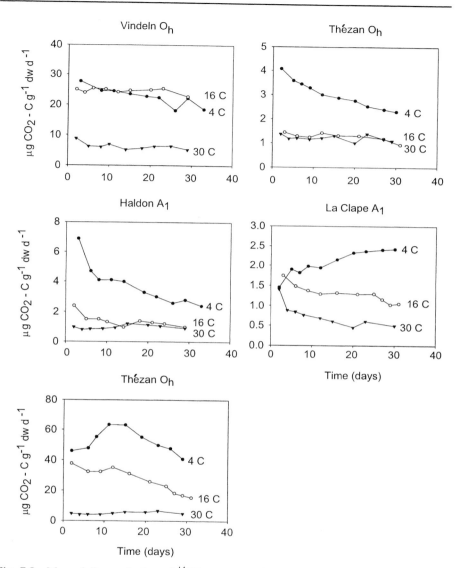

Fig. 7.8. Mean daily respiration of $^{14}CO_2$ per gram dry, partly decomposed wheat straw. Wheat straw had been incubated and was partly decomposed in the humus at five sites in a climatic transect. The straw was reincubated at 4, 16 and 30 °C in the laboratory. Decomposition was allowed to proceed until about the same mass loss in all cases (as measured from $^{14}CO_2$), the litter was reincubated at the standard temperature of 23 °C and the activity was compared among the litters incubated at different temperatures. The highest activity was found for litter that had been incubated at the lowest temperature. (Dalias et al. 2001)

The lack of a climatic influence on the decomposition of Norway spruce litter, both for the first year of incubation, and in later years, makes it differ greatly from previous studies using other litter types, notably Scots pine needles (Berg et al. 1993a,b). In other words, the decomposition of Norway spruce litter does not depend on site-specific energy and water inputs to the ecosystem but on other, more powerful influences. For Norway spruce litter, site climate based on long-term averages was very poorly related to decomposition rate, even though the variation in AET in the 1600-km-long transect ranged from 371 to 545 mm. This suggests that soil microclimate is not an important control on litter decay rates in Norway spruce stands.

For some of the Norway spruce plots in this transect, Berg et al. (1984) reported that first-year mass loss of a standardized preparation of Scots pine needles could not be correlated to climatic indices. They had incubated unified Scots pine needle litter in paired stands of Norway spruce and Scots pine, and although decomposition of the litter incubated in the Scots pine stands was regulated primarily by climate, the decomposition of that incubated in the nearby stands of Norway spruce, on the same soil and under the same climate, could not be related to any climatic factor.

Soil microclimate in spruce forests is poorly described by local temperature, precipitation and water-balance variables. Spruce trees produce dense canopies, but in a transect study Berg et al. (2000) found no effect of canopy cover and basal area indices on litter decay. In contrast, the decomposition of Scots pine litter incubated in a pine stand follows ground microclimate fluctuations very well (Jansson and Berg 1985). The results suggest that other factors may be involved, such as substrate quality and possibly different microflora in spruce, as compared to Scots pine stands. Ground climate in the spruce forests may not be related as closely to macroclimatic factors and averages as in the adjacent pine forest (Johansson et al. 1995). Under the dense spruce canopies, water could be limited due to interception, in which case temperature differences would have little effect. This appears to be a reasonable conclusion since decomposition of Scots pine needles in spruce stands was also unrelated to climate.

Dead Norway spruce needles may stay on the branches for long periods and become partly decomposed before being shed. This means that the early phase of decomposition (Berg and Staaf 1980a) occurs before litter fall, and that at least part of the litter was already in a late phase of decomposition when collected. Hence, the concentrations of compounds such as lignin, N and P will be higher than in directly shed litter and concentrations of water solubles will be lower. Further, mobile nutrients, such as K, will be leached, resulting in lower concentrations (Laskowski et al. 1995). Thus, a dominant influence of the substrate cannot be excluded. However, only one out of eight substrate-quality factors, namely initial Mn concentration, correlated positively with first-year mass loss (R^2=0.325, $p<0.05$; Table 7.8, Fig. 6.5). The relationship between Mn concentration and first-year mass loss is based on a causal relationship for the role of Mn as a rate-stimulating agent for lignin degradation (cf. Chaps. 2 and 6).

Table 7.8. Linear relationships of first-year mass loss of Norway spruce needle litter to climatic and substrate-quality factors. The litter, collected locally, was incubated in a climate transect (transect 4) ranging from the Arctic Circle in Sweden to the latitude of Copenhagen ($n=14$). All climate and substrate quality variables listed in Table 7.3 were tested

Variables	R	R^2	$p<$
MESE	0.570	0.325	0.05
AET, MESE, MAGN	0.644	0.415	0.05

7.5.2 Late stage

Individual sites. In transect 4, Berg et al. (2000) compared annual mass loss for Norway spruce needle litter to current litter lignin concentrations in the same way as was done for Scots pine needle litter (cf. Fig. 7.6). We can compare these two approaches using slope coefficients for the negative relationship between changes in lignin concentration with time, and annual mass loss. For Norway spruce needle litter, the lignin concentration at the start of each one-year period was regressed against the mass loss over that one-year period, to obtain a slope for each of the 14 sites, describing the effect of lignin concentration on litter mass loss. The values included were clearly those of the late stage (years 2, 3, 4, and 5). Lignin concentration correlated negatively with litter decay rate for 7 of the 14 sites. Berg et al. (2000) combined these into one group (Group 1, Table 7.9). For the 7 other sites (Group 2), there was no such effect related to lignin (Table 7.9).

The significant, negative slopes (Group 1, Table 7.9) for annual mass loss *vs.* lignin concentration were tested for relationships to climate and other substrate quality variables. None of the tested climatic factors (Table 7.3) was significantly related. However, initial lignin concentration gave a significant and negative relationship with slope coefficient ($R^2=0.677$; Table 7.10), and the best fit (positive) was found for the empirical relationship between litter Ca concentration and slope coefficient ($R^2=0.895$). This indicates that the higher the initial concentration of Ca, and the higher the slope coefficient, the lower the effect of lignin on litter decay rates.

Groupwise combination of data. Berg et al. (2000) performed another set of regressions, comparing the data of Group 1 and Group 2 with regard to substrate-quality variables, namely concentrations of lignin, N, Ca, Mg, Mn and water solubles (Table 6.5). For the Group 1 litter ($n=55$) they found highly significant, negative relationships between the annual mass-loss and concentrations of lignin and N (Fig. 6.9), and positive relationships for other variables including concentrations of water solubles. That water solubles correlated positively with litter mass-loss rate suggests that energy was a limiting factor.

For Group 2 ($n=33$) they compared annual mass loss and substrate quality variables (above) as potential rate-regulating factors. The relationship to Mn concentration was highly significant ($R^2=0.277$, $p<0.01$; Fig. 6.9), but there were no significant relationships between annual mass loss and the other substrate-quality factors. In both groups the interval for lignin concentrations were similar, with 227 to 524 mg g^{-1} for Group 1 and 286 to 513 mg g^{-1} for Group 2.

The litter in Group 2 had a wide range in Mn concentrations (0.41 to 7.7 mg g^{-1}; Fig 6.9). For Group 1 the range was clearly narrower (0.3-3.0 mg g^{-1}). When Berg et al. (2000) selected the corresponding range in Mn concentrations for the Group 2 litter, namely, excluded all data outside the interval 0.3 to 3.0 mg g^{-1}, no significant relationship to Mn concentration was found.

Table 7.9. Equations for the linear relationship between annual mass loss in years 2-5 and lignin concentrations at the start of each year in decomposing Norway spruce needle litter incubated at 14 sites in Scandinavia (transect 4). (Berg et al. 2000)

Site	Slope	Intercept	R	n	$p<$
Significant relationships (Group 1)					
5 Stråsan	-0.09631	55.5266	-0.709	18	0.001
111 Hässlen	-0.07393	48.2824	-0.851	13	0.001
113 Tönnersjöheden	-0.09399	57.0367	-0.973	5	0.01
10 Mästocka	-0.11077	65.05364	-0.960	5	0.01
114 Farabol	-0.10636	65.16445	-0.969	5	0.01
104 Tveten	-0.10874	66.37035	-0.955	5	0.05
102 Kungs-Husby	-0.03942	38.60125	-0.933	4	0.1
Non-significant relationships (Group 2)					
109 Ätnakobbo	0.04882	2.90427	0.801	4	*ns*
108 Västbyn	0.03572	15.61122	0.197	5	*ns*
112 Månkarbo	-0.03309	38.60966	-0.911	3	*ns*
103 Tomta	0.013382	19.7723	0.207	5	*ns*
100 Dimbo	0.037064	12.28549	0.386	5	*ns*
101 Grensholm	0.021138	12.904	0.351	4	*ns*
105 Remningtorp	0.002265	28.00804	0.063	5	*ns*

ns Not significant with p>0.1.

Table 7.10. Linear relationships between the slope coefficients from the Group 1 litter in Table 7.9 and substrate quality variables. Only statistically significant ($p<0.05$) slopes were used in this investigation. $n=7$. (Berg et al. 2000)

Equation	R	R^2	$p<$
Slope = f (Ca)	0.946	0.895	0.01
Slope = f (lignin)	-0.823	0.677	0.05

Combined data. When combining all data for late stages (Group 1 and Group 2) with the Mn concentration interval from 0.3 to 7.7 mg g^{-1} Berg et al. (2000) found a highly significant relationship between Mn concentrations and mass loss ($R^2=0.372$, $n=59$, $p<0.001$. The effects of Mn on lignin degradation have been discussed above (Chap. 3). Using all the data for late stages (Group 1 plus Group 2), including all transect data plus all data from an experimental site (B. Berg unpubl.,

$n = 95$) they found that Mn concentration correlated positively with annual mass loss ($R^2 = 0.356$, $p<0.001$; Table 6.7).

7.6 Climate and decomposition of root litter

A study of coniferous root litter was undertaken in a climatic transect across a large region ranging from the Arctic Circle ($66°N$) in Scandinavia to Berlin ($52°N$) in NE Germany. The study was carried out in coniferous monocultural forests. Berg et al. (1998) used data from 37 sites at which root litter of three coniferous species, namely, Scots pine, lodgepole pine and Norway spruce, had been incubated.

When they combined all data, the linear relationships to climatic factors and chemical composition were poor. In spite of the considerable climatic difference between sites, there were no strong relationships between any climatic variable and the first-year mass loss (range 17.0-40.9%). For the first-year mass loss, the average annual temperature (AVGT) was the most rate-regulating factor for all litter combined, with a value for R^2_{adj} of 0.186. Substrate quality indices had a weak influence. Thus, for the whole region, lignin concentration was significant with a value for R^2_{adj} of 0.142. When combining average temperature and lignin concentration the R^2_{adj} value was 0.262.

They separated pine and spruce litters and found that for the separate groups the values for R^2 increased, but still the factor average annual temperature dominated. For the pine group the R^2_{adj} reached a value of 0.346 (Table 7.11). Also over the region the N concentration in the fresh pine root litter gave a significant relationship ($R^2_{adj}=0.232$).

Table 7.11. First-year mass loss of local root litter from both Scots pine and lodgepole pine, incubated in a climatic transect, as a function of climatic and substrate quality factors. The root litter had an approximate diameter of 2-3 mm. (Berg et al. 1998)

Variable	R	R^2_{adj}	n	$p<$
Climatic factors				
AVGT	0.612	0.346	25	0.01
PET	0.563	0.287	25	0.01
AET	0.553	0.276	25	0.01
JULT	0.454	0.171	25	0.05
Substrate quality				
NITR	0.536	0.232	17	0.05
Climate and substrate quality				
AVGT, NITR	0.603	0.322	17	0.05
AVGT[a]	0.592	0.308	17	0.05

[a]Uses data for which both AVGT and NITR are available.

For the root litter of Norway spruce the average temperature in July was the strongest rate-regulating climatic factor (Table 7.12) with a value for R^2_{adj} of 0.381. A combination of temperature in July (JULT) and the initial P concentration in the litter gave an R^2_{adj} value of 0.713 for spruce root litter , thus explaining about 71% of the variation. Temperature in July and initial Ca concentration explained about 45%. Berg et al. (1998) concluded that the decomposition of spruce root litter was more dependent on energy input than that of pine, and that for both groups, energy was the main rate-regulating factor. However, fine root decay is perhaps less sensitive to temperature changes than is foliage litter decay (Silver and Miya 2001; Chap. 9).

7.7 A series of limiting factors

In later stages of decay, increasing lignin concentrations have been found to correlate negatively with lower decay rates. However, there are nutrient factors that may influence the prevailing microflora, thus influencing both the degradation rates of lignin and the substrate. The succession of the latter may be regulated by the composition of nutrients. The effects of N (Eriksson et al. 1990) and Mn (Perez and Jeffries 1992; Hatakka 2001) have been discussed earlier (Chap. 6). Lignin degradation rates as reflected through its concentration may limit litter decomposition rates if one or more of these essential elements required for microbial degradation of lignin (e.g. Mn) are limiting. At the other extreme, high concentrations of an element such as N could suppress microbial degradation of lignin. Such nutrient interactions may be complex, but the composition of the microbial community, including the lignin-degrading fungi, depends greatly on both litter degradability and concentrations of nutrient elements. If the degradation of lignin is the primary rate-regulating factor in later decomposition phases, factors such as the concentration of nutrients that influence lignin degradation will also influence the decomposition of the whole litter.

For the litter of Norway spruce, the effect of lignin was related to Mn concentrations. Within a narrow concentration interval (0.3-3.0 mg g^{-1}), Mn was not related to litter degradation rate. However, with the wider range of litter Mn concentrations (0.4-7.7 mg g^{-1}), mass-loss rates appear to be influenced by the litter concentrations of Mn and the relationship was more clear. At high Mn concentrations, microbial lignin degradation thus may be facilitated. Lignin concentration itself was less important but would increase in importance when Mn was limiting. The differing concentrations of Mn in litter could be dependent on site (soil) properties (Berg et al. 1995a), and the availability/mobility of Mn in the mineral soil could thus be an important site property for determining the rate of litter decomposition.

A similar approach may be used when discussing the effect of Ca on lignin degradation by the microbial community (cf. below). Calcium may influence the lignin-degrading microflora and thus, through lignin-degradation rates, the litter decomposition rates. The higher the Ca concentration the steeper the slope for lignin concentration *vs.* mass-loss rate ($R^2=0.895$, Table 7.10)). This is possibly due to a higher lignin degradation rate when Ca was not limiting, thus leading to less lignin regulation of litter mass loss. Early work on this relationship demonstrated that Ca concentration affects lignin decay (Lindeberg 1944).

Table 7.12. First-year mass loss of local Norway spruce root litter incubated in a climate transect as a function of some single climatic and substrate quality factors. (Berg et al. 1998)

Variable	R	R^2_{adj}	n	$p<$
Climate factors				
JULT	0.661	0.381	12	0.05
AVGT	0.588	0.281	12	0.05
AET	0.497	0.172	12	0.1
Substrate quality				
PHOS	0.569	0.239	10	0.1
CALC	0.568	0.238	10	0.1
Climate and substrate quality factors				
JULT, PHOS	0.861	0.713	12	0.01
JULT, CALC	0.713	0.454	10	0.05

7.8 Climate and the decomposition of humus and litter in humus-near stages

The idea of a low climatic influence on respiration from humus is in part supported by a study by Bringmark and Bringmark (1991) who made respiration measurements on humus in a climate transect along Sweden from the latitude of the Arctic Circle to that of the city of Copenhagen (66°08′N to 55°39′N). Incubating their samples at a standard temperature, they found higher respiration rates for the northern humus samples as compared to the southern ones. The relationship between latitude and respiration was highly significant ($R^2=0.41$; $n=166$) with respiration expressed as mg CO_2 g^{-1} ash-free humus under standardized temperature and moisture conditions. Although this measurement was made on humus, the results have a clear similarity to those of Dalias et al. (2001) described in Section 7.4.2.

Respiration rates from humus samples are sometimes negatively correlated with the N concentration of decomposing humus (Bringmark and Bringmark 1991; L. Bringmark, pers. comm.). Thus, respiration rates of humus samples collected in a transect over Sweden and kept under standard climatic conditions, showed a significant negative relationship to N concentrations in the humus ($r=-0.650$, $n=15$, $p<0.01$; Fig. 7.9; Berg and Matzner 1997). Nitrogen concentrations were in the

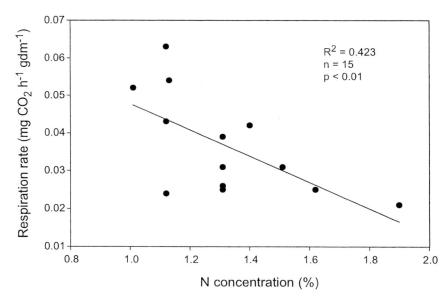

Fig. 7.9. A comparison between N concentration in humus (F and H layers) and CO_2 release rate from the same samples incubated under standard temperature and moisture conditions. Samples were collected in a transect along Sweden. (Berg and Matzner 1997)

range of 1.0-1.9%, and such a variation in N concentrations in humus could be a natural phenomenon (Berg et al. 1999b) rather than a pollution-related one. In another study, in which humus was collected from sites across Europe, from the Arctic Circle to the Mediterranean, a highly significant, negative relationship between respiration and N concentration was found (R. Laskowski, pers. comm.).

When investigating the literature on laboratory measurements of respiration from humus that has been equilibrated, we found them to be very high (B. Berg unpubl.). Thus, when recalculating amounts of CO_2 developed per hour or per day from a given ash-free sample to annual mass loss, the respiration rates often correspond to between 10 and 100% of the total humus mass per year (e.g. Persson et al. 2000), figures that do not agree with an accumulation of humus. We must conclude that such rates are not to be regarded as a quantitative measure of humus decomposition, and cannot exclude that in some cases the reported effects of rate regulating factors may be in doubt.

8 Influence of site factors other than climate

8.1 Introduction

Although litter chemical composition and climate appear to dominate as regulating factors in decomposition over a regional scale (Chaps. 4-7), there are numerous factors that are important in regulating decomposition at the local or even micro scale. These are related to soil characteristics, nutrient availability and cycling, plant community composition and structure, soil fauna, and topography. Some of these factors exert their influence by modifying the microclimate. Other factors operate primarily through biochemical or nutritional influences on microbial metabolism. Not only can these site-specific factors influence microbial metabolism, they can also alter the composition of the microbial community, as was discussed in Chapter 3.

As is the case for many ecological processes, scale is an issue. It is often possible to describe a process on a large scale with reasonable accuracy and a good understanding of the underlying mechanisms. However, when applied to a smaller scale, the regulating factors that operate at the large scale (e.g. climate) may be essentially constant, and other factors emerge.

8.2 Soil factors

Soil factors include both physical and chemical properties. Texture is perhaps the most important physical property of soil because it influences nutrient and water dynamics, porosity and permeability, and surface area. Chemical properties include pH, cation exchange capacity, and organic matter content, all of which can influence the mobility of nutrients and the composition of the microbial community. Nutrients will be discussed in Section 8.3.

8.2.1 Soil texture

One of the first long-term studies to examine the effects of soil properties on decomposition was that by Jenkinson (1977). He examined the decomposition of ^{14}C-labeled ryegrass in a variety of soils. In these soils, clay content ranged from 5 to 21%, pH from 3.7 to 8.1 and organic C from 0.97 to 4.57%. He found that neither organic matter content nor pH had much impact, except that decay was initially slower in the most acidic (pH 3.7) soil. On the other hand, soil texture was an important factor, with more litter-derived C retained in soils with higher clay content. Similarly, the total soil organic matter levels were less in sandy soils (14%) than in soils with more clay (up to 29%). This suggests that soils with clay are able to hold on to more biologically degradable SOM than are sandier soils. Several mechanisms could account for the influence of soil texture on decomposition. Jenkinson (1977) found that a double exponential model provided a better fit to long-term litter decay than did a single exponential (cf. Chap. 10).

A 2-year long field study (Perala and Alban 1982), designed to specifically test the effect of soil textures on decomposition gave somewhat different results to those of Jenkinson (1977). Their study found more rapid decay on sandier soils for all litters that they studied. The differences were rather small and it was not possible to determine the mechanisms in operation. They estimated decay rates (k) using Olson's (1963) approach in which:

$$k = \frac{L}{FF_{ss}}$$
(8.1)

where L = mean annual litter fall and FF_{ss} = steady state forest floor mass (Table 8.1). Although this method provides a useful first approach, and is much easier to use than more direct measurements of decomposition, litter fall is highly variable and a mean over 2 years may not be representative. Furthermore, forest floors, in the absence of disturbance, are not likely to be in steady state.

Table 8.1. Single exponential decay rate (k, cf. Eq. 8.1) for total litter fall of four different forests growing on loamy or sandy Alfisols in north-central Minnesota (USA). From Perala and Alban (1982)

Species	Fine sandy loam	loamy fine sand
Quaking aspen	0.17	0.18
White spruce	0.17	0.19
Red pine	0.23	0.19
Jack pine	0.19	0.20

Using very short-term laboratory studies of cellulose degradation, Schmidt and Ruschmeyer (1958) evaluated 11 different soil factors among 21 different soils by correlating them with cellulolytic activity. With regard to soil texture, they found that sand content was negatively correlated with cellulolytic activity ($r = -0.47$, $p < 0.05$). In their study, pH, ranging from 4.0 to 7.6, was the most highly significantly correlated factor for cellulolytic activity ($r = 0.76$, $p < 0.01$) and total N and nitrate-N were also positively correlated. This study is not directly applicable to field conditions, but serves to indicate the importance of soil factors including texture, pH and nutrients in decomposition processes, in this case cellulolytic activity as an index for the potential to degrade litter.

Because soil texture is so closely related to soil water dynamics, Scott et al. (1996) undertook a study to examine the effects of both soil texture and soil water pressure. They made artificial soils by blending soils collected from the field.

Their soils had sand contents of 40, 55, and 73%. In each of these soils, they regulated water at -0.012, -0.033 and -0.30 MPa. This combination of treatments yielded a continuum of water content that they described with a single variable, percentage water filled pore space. Soil texture had no effect on the decomposition of wheat litter. However, there was an interaction between soil texture and water pressure, such that the effect of water pressure was negligible in the sandy soil and was greatest in the loam soils. Also noteworthy was that the effects of soil texture and water pressure on litter decomposition were very much less than their effects on the decomposition of the older C that was already in the soil.

We conclude that soil texture is more important for long-term organic matter dynamics than for initial phases of decay. Not surprisingly, finer textured soils will interact with water more than coarser textured ones. As a result, water levels generally influence decay relatively little on sandy soils, but have a significant effect on loams or finer textured soils.

8.2.2 Forest floor type

The effect of forest floor type (humus type: mull, moder or mor) is perhaps an obvious line of inquiry. The results, however, are not as clear as might be expected. Bocock et al. (1960) incubated European ash and Durmast oak leaf litter in hairnets with 1 cm openings on mull and moder sites. Oak litter decay rates were independent of forest floor type, but ash leaves disappeared much more rapidly on mull sites (Fig. 8.1). It is important to note that there was significant earthworm (*Lumbricus terrestris* L.) activity on the mull site and that disappearance may be greater than actual decomposition because material could be easily moved out of the coarse mesh nets. A similar study compared mull and mor sites (Howard and Howard 1980). Using reciprocal transplants of hybrid oak and silver birch leaf litters, they were able to separate the effects of species and soil types over a 2-year decay period. They incubated litter in glass tubes, open to microorganisms, but closed to mesofauna including earthworms. Birch leaf litter decomposed more rapidly than that of oak, a fact attributed to the ability of birch leaf litter to retain more moisture. The impact of site (mor *versus* mull) was not consistent in their study, with birch leaf litter decomposing more slowly on mor sites, but oak leaves decomposing at the same rate on all sites.

8.2.3 Local topography

Topography, notably slope and aspect, can influence microclimate. Slope position can influence water dynamics and possibly litter accumulation. Topography is clearly important in soil formation, but its role in decomposition dynamics is not as well documented. In a forest, the canopy often moderates microclimate in the forest floor (see above) and microrelief assumes importance as a rate regulating factor.

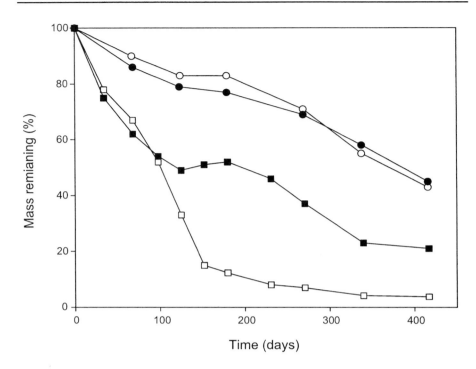

Fig. 8.1. Decomposition of deciduous leaf litters on mull, and moder forest floors. (o) oak on mull, (●) oak on moder, (□) ash on mull, (■) ash on moder (Bocock et al 1960)

Topographic heterogeneity can have a large effect on the distribution of litter on the forest floor (Dwyer and Merriam 1981). In an American beech-sugar maple forest in southern Quebec, Canada, they observed that the mass of litter accumulated on the forest floor varied from 416 g·m^{-2} on high sites to 1210 g·m^{-2} on level ground to 2438 g·m^{-2} on low sites, a factor of nearly six. The three sites also had very different levels of soil moisture and soil temperature. After 16 months, mass loss ranged from 10% on high sites to nearly 40% in low sites. Examination of bacterial populations followed the same trend, with the lowest populations on the high sites. Thus, the topography influenced the accumulation of litter and the microclimate, which influenced the microbial community and resulted in altered initial decay rates. Even though decomposition was faster at the low sites, the downslope movement of litter more than offset the enhanced decay and increased the litter accumulation. Accumulation of soil organic matter was not reported.

In many northern forests, pit and mound topography creates a micro-topography with sufficient relief to influence decomposition. The results of studies on the effects of pit and mound topography seem to contradict the results of Dwyer and Merriam (1981, above). For example, Beatty and Stone (1986) found that decomposition was slower in pits and McClellan et al. (1990) found no difference between decomposition of cellulose (filter paper) in pits and mounds. Dwyer and Merriam (1981) worked in a site that had warm, dry summers, in contrast to McClellan et al. (1990) whose sites were in cool, wet southeast Alaska where water was unlikely to be limiting. Less clear is why Beatty and Stone, working in New York State, USA, found lower decay rates in pits. Perhaps the pits in their study held water so well that the soil became hydric and the decomposition went through periods of anaerobicity.

When these studies are taken together it becomes clear that the underlying factor influencing decomposition rate is really microclimate. Whether or not microtopography influences decay rates and nutrient cycling depends on whether or not the topography causes variation in microclimate, especially moisture.

8.3 Nutrient availability

Nutrients, especially N and P, have received considerable attention in decomposition studies. The nutrient content of litter, so-called endogenous nutrients, can be a major factor in regulating the patterns and rates of decomposition (Chaps. 4-6). Somewhat less well understood is the role of nutrient availability in the environment, or exogenous nutrients. Nutrients are available to the community of decomposer organisms from natural biological and physical mineralization, from atmospheric input and from fertilization. Most of the studies on the relationship between nutrient availability and litter decay have relied on fertilization or other artificial additions. Although this approach can alter nutrient availability, there may be some impact of the pulsed nature and the timing of the nutrient addition.

Nitrogen is considered the most commonly limiting factor, and has received the most study. Adding N to decomposing substrates has been observed to increase the initial rate of decay (Tenney and Waksman 1929). Other studies with agricultural residues have found little effect of added N, except occasionally during the initial stages of decay (Lueken et al. 1962; Knapp et al. 1983; Bremer et al. 1991). However, when considering a longer period of decay, the effects of added N on decay rate appear to be negligible, and may even become negative (Fog 1988). We described mechanisms for this in Chapter 6.

To examine the influence of N availability on rates of litter decomposition, Prescott (1995) incubated lodgepole pine needle litter in forest plots that had received various fertilization treatments. These plots had significantly different amounts of available N in their soils, but there was no difference in cumulative mass loss after 33 months regardless of the available N. Prescott reported similar results, (i.e. no difference in decay rates) for litters of other forest species in response to addition of N-rich sewage sludge, and in laboratory microcosms.

In another approach to the question that relied on a naturally occurring N availability gradient, McClaugherty et al. (1985; Table 8.2) incubated six litter types in a reciprocal transplant study among five forest stands on Black Hawk Island, Wisconsin (Appendix III). The annual net N mineralization among the five stands that were studied ranged from 2.9 to 12.5 g N m^{-2} year^{-1} (Pastor et al. 1984). After 2 years of decay, the only significant differences ($p<0.05$) found were with bigtooth aspen leaves and white pine needles, both of which decomposed slower at the sugar maple stand, which had the highest N availability, than in their own stands. After 5 years of decay (McClaugherty unpubl.), white pine needles had still decomposed less in the sugar maple stand than in the white pine stand. The only other difference after 5 years was with sugar maple leaves, which had decomposed more slowly in the hemlock stand than in the other stands. The hemlock stand had the lowest N mineralization. Thus, N availability does influence the decay of litters in some, but not all circumstances.

Table 8.2. Mass lost as percent of initial after 2 and 5 years for six litter species incubated in forests with different levels of N mineralization. (McClaugherty et al. 1985, C. McClaugherty unpubl)

STAND	NMIN	Litter species					
	[$g \cdot m^{-2} year^{-1}$]	RMW	WOL	WPN	ASL	HMN	SML
2 Years							
Sugar maple	12.5	55.6	64.8	53.2	55.3	45.1	75.4
White oak	8.4	55.2	72.7				69.9
White pine	5.2	54.6		67.5			75.5
Bigtooth aspen	4.8	44.7			62.4		66.8
Canadian hemlock	2.9	38.6				47.5	66.9
5 Years							
Sugar maple	12.5	83.2	79.8	72.1	72.8	64.1	81.4
White oak	8.4	79.1	82.6				83.2
White pine	5.2	81.8		80.7			84.5
Bigtooth aspen	4.8	82.8			83.9		82.4
Canadian hemlock	2.9	82.8				64.3	74.5

NMIN N mineralization rate, *RMW* red maple wood chips, *WOL* white oak leaves, *WPN* white pine needles, *ASL* bigtooth aspen leaves, *HMN* Canadian hemlock needles, *SML* sugar maple leaves.

In contrast to mass loss, N availability does seem to have an impact on the amount of N retained (immobilized) in decaying substrates. Magill and Aber (1998), in an N enrichment study described later (Chap 12), found much more N remaining in litter on high-N treated plots. Similarly, McClaugherty (unpubl.) incubated low-N red maple wood chips and high-N sugar maple leaves across the N mineralization gradient described above (Pastor et al. 1984; Fig. 8.2). After 5 years of incubation, the concentration of N in the remaining litter was weakly ($p=0.08$) correlated with the N mineralization rate at the site of incubation.

We conclude that external N availability is not of general significance in determining rate of mass loss when considered over the continuum of litter to humus. Nitrogen availability may indirectly influence initial decay rates by influencing the nutrient content of litter, or by influencing the plant community composition, and hence the quality of the litter. However, during early stages of decay it seems that the internal nutrient composition is more important than the external. As decay moves into later stages, the external N concentration may have a positive influence on the general activity of the microbial community, but it may have a negative influence on the decay rate of the litter, due to mechanisms we have described earlier. The external N concentration may be of importance in determining the amount of N retained by decomposing litter in some systems, and hence the N storage in humus.

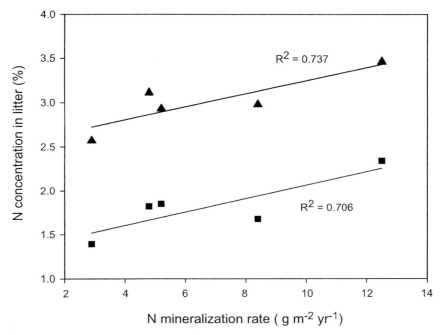

Fig. 8.2. Nitrogen concentration in sugar maple leaves (▲) and red maple wood chips (■) after 5 years of incubation in five different forests with differing N mineralization rates. All forests were located within 2 km of one another on Black Hawk Island, Wisconsin (Appendix III). From McClaugherty (unpubl.)

8.4 Plant community composition and structure

The floristic composition of a community could influence decomposition due to either synergistic or antagonistic interactions among litter types. Furthermore, the physical structure of a community could influence microclimate, with canopy cover being an obvious structural consideration. Human activities including thinning and clear cutting can be important site factors and they will be discussed in Chapter 12.

8.4.1 Effect of litter species composition

Most studies of decomposition have followed the decay of a single species of litter. A few studies have deliberately mixed litters to investigate the possibility that
a mixture reflecting the natural heterogeneity of litter fall would behave differently than a single species. Blair et al. (1990) used both single and mixed species
of foliar litter in a 1-year litterbag study. They used three common co-occurring
species that differ greatly in potential decay rates. In order of increasing resistance to decay, they were leaves of: flowering dogwood, red maple, and chestnut
oak. The individual litters decayed at rates that were expected, based on their initial chemical composition, and there was no significant effect of mixing on the decay rate. However, mixing influenced N dynamics. Addition of dogwood litter to
a mixture resulted in a greater loss of N from the mixture during the first 6 months
than was predicted based on the N dynamics of individual (unmixed) litters. However, after 1 year there were no significant differences in net immobilization of N
in the litter mixtures containing dogwood and the other litters. Unfortunately, the
study did not last into the late stage of decay, but we predict that any impacts of
mixing will diminish as decomposition proceeds.

Using a somewhat different approach, Chadwick et al. (1998) incubated Scots
pine needle litter in laboratory microcosms containing native litters from six different coniferous forests in Europe. The standard Scots pine needle litter that was
used is the same as was used by Berg et al. (1993b) in a climatic transect study.
Four of the underlying litters were Norway spruce and two were Scots pine. The
litters had a wide range of nutrient and organic chemical composition. Mass loss
of the standard Scots pine needle litter after 203 days ranged from 20 up to 38%.
Decay rate did not depend upon the type (pine *versus* spruce) of litter underlying
the litterbags, but appeared to depend on the N and Ca contents of the underlying
litter. The most rapid decay rates were found when the underlying litter had high
N and Ca contents.

8.4.2 Community structure and development

Relatively few studies have examined the importance of stand development, age
or physical structure on decomposition. Some studies have examined the role of
silvicultural practices such as thinning and clear-cutting, which will be discussed
in Chapter 12.

Many ecosystem properties change during forest succession. Turner and Long
(1975) suggested that rates of litter decomposition would change as stands develop. To test this idea, Edmonds (1979) incubated needle litter for 2 years in a
chronosequence of Douglas-fir stands ranging from 11 to 97 years of age. He also
measured air temperature, litter temperature, and litter moisture periodically
throughout the study. The youngest stand had not achieved canopy closure, but by
24 years of age the canopy was closed. The highest decomposition rate occurred in
the 24-year-old stand, where temperature and moisture conditions were most favorable for decay. The lowest decomposition rates were found in the 11-year-old

stand where the lack of canopy closure created unfavorable microclimatic conditions for decay.

In mature forests, gaps of various sizes may occur. These can have microclimatological effects that influence decomposition. In a subtropical forest in China, Zhang and Liang (1995) demonstrated that gap size could have a large effect on the rate of mass loss in leaf, bark and branch litter. They were able to demonstrate that gap size was a proximate factor with the ultimate factor being primarily soil moisture. Naturally occurring gaps ranged from 5 to 40 m in diameter. As the gap sizes increased, leaf litter annual mass loss rates decreased from 57 to 44%. Mean gravimetric soil moisture decreased from 19.2% in the closed forest to 11.4% in the largest gaps. This factor alone gives a highly significant correlation between decay rate and soil moisture (R^2=0.922, $p<0.001$). It is important to note that this study was done in a subtropical forest, and the effects of gap size may change as latitude increases, as suggested by Yin et al. (1989).

Lockaby et al. (1995) superimposed a decomposition study onto a community manipulation study in loblolly pine stands. Vegetation suppression created four different communities. The treatments were: (1) none (control); (2) deciduous woody suppression; (3) herbaceous suppression; and (4) pine only with both woody and herbaceous suppression. After 20 months, mass loss was highest in treatment plots where herbaceous vegetation had been removed, and lowest in the pine-only treatment (no. 4). The mechanism for this difference is not known.

8.6 Carbon dioxide levels

Atmospheric CO_2 continues to increase and may double by the year 2100 (Watson et al. 1992). Considerable scientific attention has been directed towards an understanding of how increased CO_2 might influence ecosystems. Accordingly, several studies have looked at the influence of CO_2 concentrations on decomposition. The influences can be direct or indirect. Direct influences would occur if changing CO_2 levels influence the microbial communities, or the biochemistry of decomposition. Direct effects are unlikely, or at least hard to document, because by its nature, decomposition occurs in a CO_2 enriched environment. One type of indirect effect would occur if changing CO_2 levels altered the chemistry of the litter. Another type of indirect effect would occur if there were climate changes. The latter effects are discussed in Chapters 7 and 12.

Franck et al. (1997) hypothesized that increased atmospheric CO_2 would cause plants to produce lower quality litter that would decompose and release nutrients more slowly. They tested this by growing four grass species at ambient and twice-ambient levels of CO_2 and following the C and N dynamics of the litter. The results were full of interactions. The litter of species grown under enriched CO_2 had higher decay rates for three species (perennial ryegrass, soft chess and wild oats) and lower for the only native grass in the study (small six weeks grass). Though not examined in the study, it is possible that the higher initial decay rates in the litter grown with CO_2 enrichment, were due to increased production and storage in litter of labile C compounds.

Cotrufo and Ineson (1995) conducted a similar laboratory microcosm study with tree roots. They grew birch and Sitka spruce at 350 and 600 ppm of CO_2, and under two nutrient regimes. Spruce root litters had the same mass loss regardless of treatment, but birch roots grown under elevated CO_2 had slightly, but significantly, lower mass losses. This may be due to changes in litter chemistry, as they reported an increase in C/N ratio in roots grown under CO_2 enrichment. In a subsequent field study (Cotrufo and Ineson 1996), birch was grown under ambient and elevated levels of CO_2. They examined both the litter chemical composition and its decomposition, and found that the lignin content of birch leaf litter increased from 17.7 to 28.7% when the atmospheric CO_2 concentration during growth increased from 350 to 600 ppm. Adding a slow release mixed fertilizer containing N, P and K decreased lignin concentrations to 13.3 and 17% at low and high CO_2, respectively. As a result, initial decay rates increased. Remarkably, unfertilized litter grown under 600 ppm CO_2 had nearly three times as much mass remaining after 1 year in the field as did fertilized and low CO_2 litters.

In a review on the effects of increased atmospheric CO_2 on litter quality and decomposition, Cotrufo et al. (1998) suggest that the major effect is an increased C/N ratio, though relatively few species have been studied. This decrease in N concentration may have different effects on the successive stages of decay (cf. Chap. 2), possibly slowing decay during early stages, but actually increasing the total amount of mass lost by the time the litter enters the humus-near stage. Elevated CO_2 may influence the composition of the plant community in ways that would also cause changes in patterns of decomposition. For example, the relative abundance of N_2-fixing plants, or the ratio of C3 to C4 plants, could change if atmospheric CO_2 levels increase or climate changes (Zanetti and Hartwig 1997; Huang et al. 2001).

9 Decomposition of fine root and woody litter

9.1 Introduction

Most studies of litter decomposition in forests have focused on foliar litters because of their large amounts and relatively high nutrient contents. Foliar litter provides an important transfer of organic matter and nutrients to the soil, and the patterns of its deposition are temporally regular and spatially rather uniform. Woody and fine (small diameter) root litter can also contribute large amounts of organic matter to forest soils. Fine root litter inputs are highly variable across ecosystems, but in at least some ecosystems, they represent a transfer of organic matter and nutrients to the soil of the same magnitude as foliar litter. Woody litter is deposited sporadically in time and space. Its deposition may be trivial in managed forests.

Fine roots are variably defined, usually based on diameter, and represent a rapidly changing component of forest biomass. Fine root litter differs from woody litter in several key respects. Fine roots may comprise less than 2% of the biomass present in a forest ecosystem, but they may contribute as much as 40% of the annual production (Vogt et al. 1990). Fine roots and green leaves are among the most nutrient rich of plant tissues (Table 9.1). Furthermore, they have one of the highest surface area-to-volume ratios of any litter type. With all of these characteristics we would expect fine roots to decay very rapidly, but that is not the case as will be discussed below.

According to Harmon et al. (1986), woody litter plays important but poorly studied roles in forest ecosystems. Woody litter includes stems, stumps, branches, twigs and most roots, excluding those with a small diameter. These litter components enter the soil very erratically in space and time, often as a result of such sporadic events as strong wind, heavy snowfall, or freezing rain. Further, the tendency for the amount of woody litter fall to increase with stand age, gives it a different pattern as compared to foliar litter. A single tree may fall as a result of a storm or death. The bulk of its biomass is concentrated along the bole and branches. The chemical and physical quality of this woody litter is quite different from foliar litters.

The amount of woody litter present in forests varies by two orders of magnitude from about 1 Mg ha^{-1} in dry tropical forests to 500 Mg ha^{-1} in old-growth coniferous forests in the Olympic Mountains in the Pacific Northwest of the United States (Agee and Huff 1987). However, most ecosystems fall into the range of 5 to 50 Mg ha^{-1} (Table 9.2). Woody litter often increases immediately after a harvest due to logging slash (McCarthy and Bailey 1997; Table 9.2). However, older managed stands can have less woody litter than do corresponding old-growth stands (Goodburn and Lorimer 1998; Table 9.2). Estimates have been made about the proportion of woody litter fall to foliar litter. Using the foliar litter fall as a reference in a mature Scots pine forest, the woody litter fall (cones excluded) made up ca. 10% and cones separately, ca. 25% (Berg et al. 1993d).

Table 9.1. Representative concentrations of nutrients in living leaves, wood and fine roots. For leaves and wood, values are averages for four species groups: European white birch, common oak, filbert, and European ash. For fine roots values are taken from a northern hardwood forest containing sugar maple, yellow birch, American beech, and red spruce

	Concentrations as % of dry matter				
	N	P	K	Ca	Mg
Leaves[a]	2.28	0.19	1.45	1.26	0.29
Wood[a]	0.30	0.03	0.21	0.32	0.046
Fine roots[b] (0.6 mm)	2.00	0.10	0.23	0.21	0.05

[a]Swift (1977).
[b]Fahey et al. (1988).

Not only is woody litter deposited unevenly, it comes in a huge array of sizes and conditions. Clearly the diameter of the woody debris will influence surface area-to-volume relationships in a way that results in more rapid decay of smaller diameter pieces. Furthermore, wood can undergo extensive decay before falling to the ground. This could be due to pathogens on living trees or saprophytic organisms on snags. In contrast, a storm may blow over or break limbs from a living tree.

One of the most important aspects of woody litter in terms of its decomposition is its extremely low nutrient content. This means that organisms that consume wood must either consume very large quantities in order to extract sufficient nutrients, or that the nutrients must come from outside the substrate. Such a low nutrient environment would be most suitable for organisms with low nutrient demands. A comparison of initial nutrient concentrations in wood, fine roots and leaves is shown in Table 9.1.

9.2 Woody litter decomposition

9.2.1 Methods

Decay classes for coarse wood (logs)

The state of decay of logs is often categorized using decay classes. Decay classes are based on visual and physical properties of wood. Although different investigators have given somewhat different definitions to each decay class, this approach is widely used, and a general scheme for decay classes is given in Table 9.3. Much of the recent literature has used a five-stage progression of decay, designated as

Table 9.2. Mass of dead wood on the ground in selected types of forest stands

Forest type	Location	Mass [Mg ha^{-1}]
CONIFEROUS STANDS - BOREAL AND TEMPERATE		
Douglas-fir - hemlock[a]	Oregon/Washington, USA	500
Boreal fir-spruce[b]	Newfoundland, Canada	4-22
Rocky Mt. spruce-fir[c]	Colorado, USA	52
Ponderosa pine[d]	Colorado, USA	3-5
DECIDUOUS STANDS - BOREAL AND TEMPERATE		
Aspen[e]	Alberta, Canada	15-25
Southern maple[f]	Tennessee, USA	14
Northern hardwood[g]	New Hampshire, USA	21-30
Hemlock-hardwoods[h]	Wisconsin/Michigan, USA	16
Mixed oak[i]	Kentucky, USA	16-22
Southern beech[j]	New Zealand	300
MANAGED TEMPERATE STANDS		
Northern hardwood - even-aged[k]	Wisconsin/Michigan, USA	6
Northern hardwood - selection[k]	Wisconsin/Michigan, USA	15
Northern hardwood - old-growth[k]	Wisconsin/Michigan, USA	29
Appalachian hardwood 2 years[l]	Maryland, USA	55
Appalachian hardwood 25 years[l]	Maryland, USA	17
Appalachian hardwood 80 years[l]	Maryland, USA	19
Appalachian hardwood >100 years[l]	Maryland, USA	33
TROPICAL STANDS		
Tropical thorn woodland[m]	Venezuela	1
Tropical very dry[m]	Venezuela	1
Tropical transition[m]	Venezuela	3
Tropical moist[m]	Venezuela	18
Tropical low montane moist[m]	Venezuela	21
Tropical montane wet[m]	Venezuela	18

[a]Agee and Huff (1987).
[b]Sturtevant et al. (1997).
[c]Arthur and Fahey (1990).
[d]Robertson and Bowser (1999).
[e]Lee et al. (1997).
[f]Onega and Eickmeier (1991).
[g]Gore and Patterson (1986).
[h]Goodburn and Lorimer (1998).
[i]Tyrell and Crow (1994).
[j]McCarthy and Bailey (1997).
[k]Muller and Liu (1991).
[l]Stewart and Burrows (1994).
[m]Delaney et al. (1998).

decay classes I through V. This concept was first articulated by Maser et al. (1979) for coniferous trees, and has been recently adapted to deciduous trees by Pyle and Brown (1998). One of the problems with this system is that objects as large as logs do not decay uniformly. First, logs are composed of different quality substrates (e.g. inner and outer bark, sapwood, heartwood) each of which decays at a different rate, and begins to decay after different lag periods (Schowalter 1992). Furthermore, a particular log may be invaded by fungi with different mechanisms for decomposing wood (Boddy et al. 1989), and thus may contain parts in several different decay classes (Pyle and Brown 1999).

Mass loss rates: percent loss and decay constants (k)

Many reports of wood decay use a decay constant, calculated in the sense of Olson (1963), see Eq. (10.1). It can be calculated based on a single point, using Eq. (9.1).

$$
k = \frac{-\ln(\frac{M_t}{M_0})}{t} \tag{9.1}
$$

where M_t is mass at time, t and M_0 is initial mass ($t=0$).

The decay rate constant, k, can also be calculated using multiple points over time and fitting a linear regression to the line formed by the natural log of mass *versus* time. The slope of this regression line is $-k$. This latter approach is better than the single point approach because it includes data taken over the course of decay. However, most litters, and especially wood, seldom decay at a constant rate and the use of k can thus be misleading. For example, Fig. 9.1 shows both percentage mass loss and the changing value of k during decomposition of logs. Whenever available, we present decomposition data as a percentage of the initial mass that is lost; however, many studies present k values and we report those also.

Table 9.3. System of decay classes used in wood decay studies. Modif ied from Maser et al. (1979) and Pyle and Brown (1999)

Characteristic	Decay class				
	I	II	III	IV	V
Bark attached tightly	+	-	-	-	-
Wood not stained	+	-	-	-	-
Bark present, perhaps loose	+	±	-	-	-
Twigs retained	+	±	-	-	-
Wood solid, resistant	-	+	±	-	-
Log surface may flake, fall into shreds	-	-	+	-	-
Log solid but decay clearly evident	-	-	±	+	-
Logs easily broken into large pieces	-	-	-	+	-
Log easily crushed	-	-	-	+	±
Log more than 85% powdery	-	-	-	±	+
Log shape oval to nearly flat	-	-	-	±	+

+ Present; ± present or absent; - absent.

Estimating mass loss in coarse woody litter

Litterbags have been the dominant technique for studying foliar litter decomposition. For coarse wood, the sheer size and the long life expectancy of the log make litterbags inappropriate. Most studies have relied on density changes to estimate mass loss, though this approach is not without its own problems. Volume displacement in water as a means of determining bulk density is commonly used, but it becomes less accurate as wood enters later stages of decay and loses its structural integrity. Further, bulk density measures do not account for losses due to fragmentation and removal by tunneling arthropods. The methods for studying woody detritus have been well-summarized by Harmon and Sexton (1996).

Litterbags have been used to study the decay of wood chips (see below). This technique allows for a more direct comparison of wood with foliar litter, as a substrate, by eliminating the differences caused by the volume of the log.

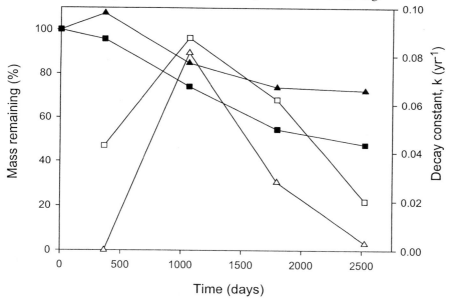

Fig. 9.1. Percentage of initial mass remaining and decay rate constants (k, sensu Olson 1963) during each measured period for red pine and red maple logs incubated under a temperate climate in a red pine plantation and an adjacent mixed hardwood forest at Harvard Forest, Massachusetts, USA. k-values were calculated for each interval of the study (■) Red maple % mass remaining, (▲) red pine % mass remaining, (□) red maple k, (△) red pine k. (C. McClaugherty unpubl.)

9.2.2 Decomposition rates *versus* climate

Mass loss rates of woody litter vary dramatically among climatic zones (Table 9.4) and among litter species. Yin (1999) has presented a very thorough compilation of woody debris decay studies. Temperature appears to be the most important climatic variable for large logs, at least in a wide range of tropical systems (Chambers et al. 2000), perhaps because the volume of the logs makes them able to buffer changes in moisture. In fact, too much moisture could suppress log decay by creating zones of low oxygen availability. However, the variation in decay rates among species within a single climate, can be greater than the variation of a single species across a range of different climates. The differences can generally be attributed to substrate quality, though this term is somewhat hard to define for wood.

Table 9.4. Representative decomposition rate constants (k, year^{-1}, see Eq. 9.1) for large woody debris

Ecosystem	Location	k
Boreal coniferous[a]	S Norway	0.033
Boreal coniferous[b]	NW Russia	0.019-0.108
Temperate deciduous[c]	Tennessee, USA	0.086
Temperate deciduous[d]	Indiana, USA	0.018-0.045
Temperate deciduous[e]	New Hampshire, USA	0.096
Temperate mixed[f]	Minnesota, USA	0.042-0.080
Temperate mixed[g]	Michigan/Wisconsin, USA	0.021
Temperate coniferous[h]	British Coloumbia, Canada	0.022
Tropical evergreen[i]	Brazil	0.015-0.67
Tropical dry[j]	Mexico	0.008-0.615

[a]Næsset (1999).
[b]Harmon et al. (2000).
[c]Onega and Eikmeier (1991).
[d]MacMillan (1988).
[e]Arthur et al. (1993).
[f]Alban and Pastor (1993).
[g]Tyrell and Crow (1994).
[h]Stone et al. (1998).
[i]Chambers et al. (2000).
[j]Harmon et al. (1995).

It is possible that physical factors relating to wood structure, such as porosity, play a more important role in regulating decomposition than do different nutrient levels. Thus, in an experiment with aspen and European beech wood sticks incubated for a year in the forest floor, the more porous aspen wood lost more than 50% of its initial mass, while the less porous beech had a barely measurable mass loss (B. Berg unpubl.).

Mass loss rates change over time as logs become more decayed. McClaugherty (unpubl.) has followed decay of red maple and red pine logs for 7 years in a temperate deciduous forest and an adjacent red pine plantation at the Harvard Forest (Appendix III, Fig. 9.1). The species decay at different rates, but both have similar patterns of decay. The decay during the first year was very slow (not measurable in red pine). The highest decay rates occurred in the interval between years 1 and 3. The decay rate then slows in the subsequent periods. Schowalter et al. (1998)

observed a similar phenomenon in oak logs, describing their decay with a two-phase exponential decay model. They suggested that a slower, third exponential decay phase would likely emerge as decomposition proceeded.

The kinetics of decay in coarse woody debris is complicated because different parts of the log may be undergoing attack by microbial species or communities with very different metabolic abilities. In the simplest sense, an entire log could be invaded by either white-rot or brown-rot fungi.

Smaller diameter woody debris has been less well studied. Erickson et al. (1985) compared the decay of logging residue in two diameter classes in a variety of temperate coniferous forest ecosystems in Washington State, USA. They found that smaller diameter (1-2 cm) twigs decayed much slower than did the larger diameter (8-12 cm) pieces, and attributed this to the fact that smaller diameter fragments dried more quickly, thus suppressing decay. In their study, decay rate constants (k) for small diameter slash ranged from 0.004 to 0.011, much lower than for most large or coarse woody debris (Table 9.4).

Twigs (<5 cm diameter) of red pine and red maple, tethered to nylon strings on the forest floors of a red pine plantation and a red maple - red oak forest, respectively, in Massachusetts, exhibited highly variable mass loss (Fig. 9.2, McClaugherty unpubl.). After 5 years, red pine twigs had lost considerably less mass (35.0%) than had those of red maple (74.6%). Diameter had no effect on mass loss from red maple twigs. A linear regression for pine twig mass loss *versus* diameter was significant ($p<0.001$), but the R^2 was only 0.266.

9.2.3 Carbon dioxide release

Decomposition of wood has also been measured as CO_2 release or respiration. This method allows for more instantaneous assessment of C flux as compared to periodic mass loss measurements. In a pair of studies, Marra and Edwards (1994, 1996) measured respiration from logs in decay classes I through III and V (they did not measure decay class IV), in a clear-cut and an old-growth forest on the Olympic Peninsula of Washington State, USA. As expected in this seasonal environment, respiration reached a maximum in summer and a minimum in winter. Variability was greater in the clear cut, but overall there was no difference between the clear-cut and old-growth forest. Species did differ, however, with western hemlock logs having higher respiration rates than those of Douglas-fir. This reflects the greater amount of inhibitory secondary compounds such as tannins and extractable phenols found in the wood of Douglas-fir (Kelsey and Harmon 1989). The results are similar to other studies that have compared the decay of these two species (Graham and Cromack 1982).

9.2.4 Organic chemical changes

Studies of chemical changes during litter decay began in the early 20th century. Among the early studies, Rose and Lisse (1916) described the chemical composition of fresh, partially decayed and "completely" decayed Douglas-fir wood. "Completely" decayed means that the wood has lost its structural integrity and would correspond to decay class V (Table 9.3). They noted the resistance

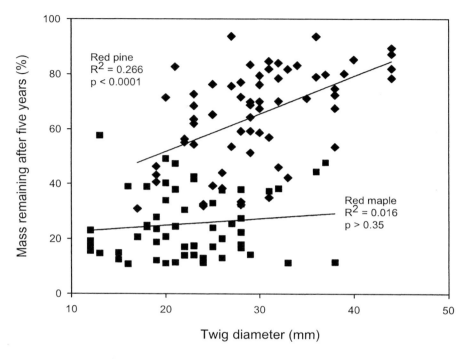

Fig. 9.2. Mass remaining after 5 years of field incubation of twigs (<5.0 cm diameter). Twigs were incubated in a mature red pine plantation or in a mixed mesophytic forest dominated by red oak and red maple. Study was conducted at the Harvard Forest (see Appendix III). (♦) Red pine, (■) red maple. (C. McClaugherty unpubl.)

of lignin to decomposition and suggested the possibility of connecting decayed organic matter and lignin residues to humic substances in the soil. Bray and Andrews (1924) demonstrated that different fungi acted differently on the cellulose and lignin of wood (cf. Chap. 3), with brown-rots acting only on cellulose. Using more recent technology, namely ^{13}C nuclear magnetic resonance, Preston et al. (1990) characterized the chemical changes in decaying heartwood of Douglas-fir, western hemlock and western red cedar. They distinguished changes in the relative amounts of C in carbohydrates, lignin, aliphatic groups and the sum of carboxyl plus carbonyl groups. Douglas-fir and hemlock wood followed the more classic pattern, with a rapid decline in carbohydrate-C to less than 10% of original amount, and an increase in the relative amount of lignin C. In contrast, western red cedar wood showed little change in its organic chemical composition, even though density had declined and the physical structure had collapsed.

The loss of holocellulose dominates mass loss, at least during the initial stages of wood decay. Figure 9.3 shows the amounts of lignocellulose, lignin and extractives remaining in wood chips of red maple and white pine during 42 months of decay on the forest floor of Black Hawk Island, Wisconsin (Appendix III). Although red maple wood lost considerably more mass than that of white pine, both species exhibited similar patterns of mass loss and change in chemical composition. Mass loss patterns were similar to those observed earlier: slow during the first year, faster during the second year, and then slower again in the third year.

9.2.5 Changes in nutrient concentrations

Woody litter contains relatively small amounts of nutrients, especially when compared to foliar and root litter (Table 9.1). However, there is no clear agreement on the nature of nutrient dynamics in woody debris. Most investigators have found that decomposing wood accumulates some elements (e.g. N, P), retains some, and releases still others (e.g. K). Alban and Pastor (1993) studied logs of aspen, spruce, red pine and jack pine that had decayed for 11 to 17 years, and compared the amounts and concentrations of nutrients in the original logs with those in the decayed logs. The concentration of N increased by an average factor of 4.3, from 1090 to 4698 ppm. Concentrations of P also increased in all four species, with the average increase from 120 to 348 ppm, a factor of 2.9. Potassium concentrations remained about the same, with the average concentration increasing from 731 to 784 ppm. Concentrations of Ca increased from 2225 to 5784 ppm and Mg increased from 206 to 504 ppm. Initial concentrations of all nutrients were higher in aspen wood than in that of the conifers, but the nutrient concentrations increased by a larger factor in the three conifer species. In a study of Sitka spruce and western hemlock, Graham and Cromack (1982) found that N accumulated (net transport into the wood) in western hemlock, but not in Sitka spruce, though the concentration of N increased in both species. Concentrations of other elements (P, Ca, Mg, K, Na) did increase, but their amountsdid not increase.

When comparing net movement of nutrients over the entire span of decomposition (all five decay classes) for Douglas-fir wood, Means et al. (1992) observed that N and K were lost, Ca and Mg were accumulated, and P and Na were accumulated and then released, with no overall net change. In a northern hardwood forest, Arthur et al. (1993) examined the change in nutrients in logs that had been on the ground for 23 years following a clear-cut. They estimated that during that time the logs had released, as a percentage of initial content, 31% of N, 68% of P, 77% of Mn and between 86 and 93% of Ca, K, Mg, and Na. At different stages of decay and for different nutrients, logs may be either net sinks or net sources of nutrients for the soil.

Krankina et al. (1999) measured the concentrations of 12 nutrients (N, P, K, Ca, Mg, Al, Cu, Fe, Zn, Na, B, Mn) in decaying logs of Scots pine, Norway spruce and birch in northwestern Russia. They found that nearly all nutrients showed an increase in concentration as logs passed through decay classes III, IV and V. The exceptions were K, which did not increase in pine and birch logs, and B, which did not increase in spruce and pine logs. Although concentrations of many of the elements increased, there was no net accumulation of nutrients over the course of decay, except for Al, which accumulated in wood of pine and birch, and Na, which was retained throughout decay.

N_2-fixation. With the low N concentration in wood, even low N_2-fixation rates could be important in influencing the N concentration of the wood during its decay. N_2-fixation has been noted in decaying logs, and may be of importance in the N dynamics of wood decay. Using acetylene reduction to estimate the potential for N_2-fixation, Larsen et al. (1978) found that brown-rotted wood was more likely than white-rotted wood to support N_2-fixation, and that there were differences between species, with Douglas-fir wood having higher rates. Jurgensen et al. (1984) and Griffiths et al. (1993) found that the N_2-fixation potential increased as decay proceeded. The amount of N_2-fixation in rotting wood is generally low, but potentially important to the N dynamics of the log (Larsen et al. 1982).

Fungal transport. Fungi that decompose wood can be important in transporting nutrients into or out of the decaying wood. Although the actual work of decay is done by hyphae penetrating the wood, nutrients are transferred to, and accumulated by, two particular fungal structures that are on or above the surface of the wood: rhizomorphs and sporocarps. Sporocarps are the fruiting bodies or reproductive structures of fungi and they often appear on wood during the first decade of its decay. Their tissues are greatly enriched in nutrients compared to their woody substrate. Harmon et al. (1994) found that fungi had concentrations of N, P and K that were 38, 136 and 115 times, respectively, as great as the concentrations in the logs on which the fungi were growing. This enrichment occurs largely by mycelial transport from the log and the surrounding environment into the sporocarp. Although small in mass, the fungi transferred measurable amounts of nutrients out of the logs and into the sporocarps, clearly against the nutrient gradient. The amounts of nutrients found in the sporocarps as a percentage of the total amounts initially present in the logs were 0.9-2.9% for N, 1.9-6.6% for P and 1.8-4.5% for K.

After fungi have decayed a substrate, they must eventually forage for new substrates, reproduce sexually, or perish. In many basidiomycetes, fungal hyphae can aggregate into cords or rhizomorphs and grow until they encounter suitable substrate. The cords can form long-lived networks that have the ability to transport e.g. C and P over one meter (Boddy and Watkinson 1995). For both fruiting bodies and rhizomorphs, nutrients are moved and concentrated.

Although wood is low in nutrients, it demonstrates a wide variety of decomposition patterns. Because of its low nutrient status, its decay may be more dependent on the exogenous supply of nutrients. Given this wide diversity of nutrient dynamics, it is not at present possible to define a general conceptual model for changes in nutrient concentrations during wood decay. The variability among the example studies given above may be due to species, environmental and methodological differences.

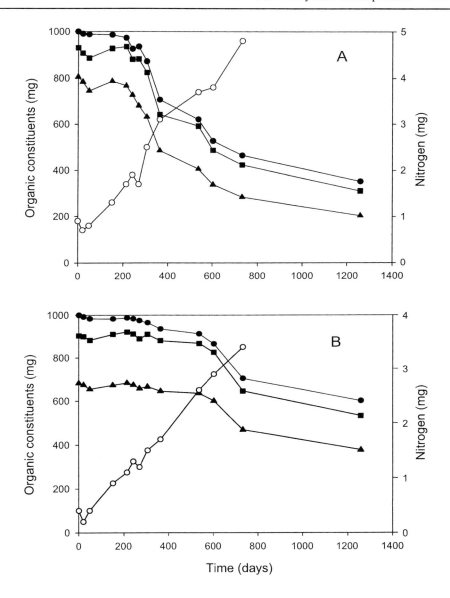

Fig. 9.3. Changes in amounts of some organic constituents and N over time in wood chips incubated in litterbags on the forest floor of a sugar maple forest on Black Hawk Island, (see Appendix III). Organic constituents are expressed as mg g^{-1} of initial material. Nitrogen expressed in mg g^{-1} of initial material.. **A** Red maple wood. **B** White pine wood. (●) Total organic matter; (■) lignocellulose; (▲) Lignin; (○) total N. (C. McClaugherty unpubl.)

9.3 Fine root decomposition

Estimates of root litter input are highly variable. Because of the difficulty of measuring root processes, few studies on fine root litter production have been done. Similarly, root decomposition is below ground and difficult to examine without significantly disturbing the system. Nevertheless, fine root litter may represent a large input of organic matter into ecosystems, and the decay of roots has implications for organic matter and nutrient dynamics in ecosystems.

It is important to note that fine roots are markedly different from larger diameter roots. The definition of fine roots cannot be made solely on the basis of diameter because species differ. Fine roots can be defined to include root tips and small diameter roots without secondary growth. Thus, fine roots have, in general, about the same diameter or less than root tips. In fact, root tips may be slightly larger in diameter due to mycorrhizal coverings.

9.3.1 Fine root litter

Amounts of litter

In contrast to aboveground litter, it is not possible to directly measure fine root litter production. A variety of methods have been used to approach this problem. Several investigators have used sequential measurements of live and dead roots to develop estimates of production, mortality and decomposition (Persson 1980; McClaugherty et al. 1982; Santantonio and Hermann 1985). This mass balance approach assumes that decreases in living biomass are due to death, which is therefore litter production. Timing and frequency of sampling are important in this technique. Kurz and Kimmins (1987) analyzed this method using a computer simulation and found that the estimates were very sensitive to violations of the assumptions and that estimates could be either too high or too low.

Rhizotrons (direct viewing through underground windows or tubes, Hendrick and Pregitzer 1992; Burke and Raynal 1994) allow for direct observation, but some disturbance of the rooting environment is inevitable. Another method involving elemental budgets was proposed by Nadelhoffer et al. (1985), and developed further by Raich and Nadelhoffer (1989). In this technique, fine root production is estimated by using fine roots to "balance" the nutrient budget.

With all these techniques (see review by Hendricks et al. 1993), we are still not able to precisely measure the transfer of fine-root litter to the soil. Vogt et al. (1996) reviewed the literature and found 41 data-sets that included estimates on below-ground litter transfer. The values were derived from different investigations, and the methods were not comparable. Nevertheless, the estimates for below-ground litter input, which is predominantly from fine roots, ranged from 100 g m^{-2} year^{-1} in a northern hardwood forest in New Hampshire, USA, to 1262 g m^{-2} year^{-1} in a Pacific silver fir forest in Washington State, USA, with a mean of 436 g m^{-2} year^{-1}. These values are similar to the amounts of foliar litter fall.

Chemical composition of fine roots

Fine roots have rather high "lignin" concentrations as compared to wood and foliage (Table 4.8), in the range of 25 to 50% (Vogt et al. 1991). Nitrogen concentrations also tend to be relatively high, generally in the range of 1 to 2%, especially as compared to wood (Table 9.1), but other nutrients are less predictable. Because of their location in the soil, fine roots may accumulate elements such as aluminum (Dahlgren et al. 1991). Furthermore, the tips of many, if not most, fine roots in forests are mycorrhizal. The mycorrhizal association may have a direct influence on chemical composition due to the presence of fungal biomass, as well as indirectly, by influencing nutrient concentrations in the root environment, and possibly influencing the decay resistance of the root (Harley and Smith 1983).

9.3.2 Mass loss rates

Researchers have had considerable difficulty reconciling the apparent high productivity of fine roots in forest ecosystems with the apparent low decomposition rate of fine root litter in forests (Fahey and Hughes 1994). It is unclear whether the problem is due to errors in the production estimates, or the decomposition estimates, or both.

Litterbag methodology. Most studies of mass loss in fine roots have utilized a litterbag approach, though some have used sequential measures of dead and live root mass (see above). Litterbags are generally filled with excised live roots that are dried, killing at least a proportion of the attached mycelia. For foliar litter, the bags are generally filled with newly senesced material. Another difference in comparison with foliar litter is that fine root litter remains in the place where the roots die, possibly already attached to, or at least closely surrounded by saprophytic microorganisms. Removing fine roots from this environment and placing them into a litterbag and reburying them is much more drastic than placing leaves or needles into litterbags. Nevertheless, litterbags continue to be used and, so far, there have been no definitive studies that show if or how much litterbags modify the decay of fine roots. Realizing that fine root decay studies are subject to a rather large and uncertain amount of error, we will nevertheless proceed to summarize some of these studies.

Studies on fine root mass loss. Selected first year mass losses (%) for a variety of fine roots are given in Table 9.5. In boreal forests, first year mass loss ranged from 19 to 40%, with much of the variability due to climate (Berg et al. 1998). One study from Puerto Rico compared two species, Sierra palm and Tabonuco, a tropical hardwood tree. The roots from the hardwood tree almost completely disappeared after 1 year while the palm roots lost only 37 to 43% (Bloomfield et al. 1993). In this case, climate was similar, so the differences between the two species were most likely due to differences in chemical composition.

Table 9.5. First year mass loss of fine roots in a variety of ecosystems

Species/ System type	Diameter [mm]	Mass loss 1st year [%]	Comments
BOREAL			
Scots pine[a]	2-3	19.3-40.9	Sweden, latitudinal climatic transect, 18 sites
Norway spruce[a]	2-3	27.4-39.7	Sweden, latitudinal climatic transect, 12 sites
Lodgepole pine[a]	2-3	29.8-35.3	Sweden, 4 sites
Norway spruce[b]	<1	23.5	Estonia
Norway spruce[b]	1-2	21.6	Estonia
TEMPERATE			
N. hardwoods[c]	<1	17	White Mountains, New Hampshire, USA
N. hardwoods[d]	<0.5	30	Adirondacks, New York, USA
N. hardwoods[d]	0.5-1.5	17.4	Adirondacks, New York, USA
N.hardwoods[d]	1.5-3.0	20.9	Adirondacks, New York, USA
White pine[e]	<3	21.5	Sugar maple forest, Black Hawk Island
Sugar maple[e]	<3	15.4	Sugar maple forest, Black Hawk Island
TROPICAL			
Tabonuco[f]	<2	99.8-99.9	Puerto Rico, tropical montane rain forest
Sierra palm[f]	<2	36.7-43.4	Puerto Rico, tropical montane rain forest

[a]Berg et al. (1998).
[b]Lõhmus and Ivask (1995).
[c] Fahey et al. (1988), northern hardwoods mean for sugar maple, American beech, yellow birch, and red spruce.
[d]Burke and Raynal (1994), northern hardwoods sample from forest dominated by sugar maple, American beech, yellow birch, and red maple.
[e]Aber et al. (1984) site description in Appendix III.
[f]Bloomfield et al. (1993).

Decomposition of fine roots of sugar maple and white pine roots was followed in a 10-year litterbag study in a temperate sugar maple forest on Black Hawk Island, Wisconsin (Appendix III; McClaugherty unpubl.; Fig. 9.4). After 10 years, sugar maple fine roots had lost 56.5% of their initial mass and white pine fine roots had lost 66.4%. White pine roots lost more mass than sugar maple roots during their first year of decay. After that, decay rates remained constant. This illustrates again that relying on first year mass loss rates to predict longer term decomposition may be incorrect.

Factors that influence fine root decay rates. Silver and Miya (2001) collected and analyzed 176 data sets on root decomposition. Their analysis included broadleaf, coniferous and graminoid roots, and sites with latitudes ranging from 4°N to 66°N. Using a stepwise multiple regression they found that AET, root Ca concentrations and C-to-N ratios accounted for 90% of the variability in early stage root decay rates (k-values). They concluded that, on a broad scale, root chemistry was the primary determinant of root decomposition rates, with secondary roles for climate and environmental factors. Looking at fine roots only (defined here as < 2mm in diameter), conifer roots decayed much more slowly than broadleaf roots. It is noteworthy that their results show a greater importance of chemistry than climate, which is in contrast to what has been observed for most foliar litter species (cf. Chap. 7). They suggest that because roots are buried in the soil, they are more buffered from climatic conditions.

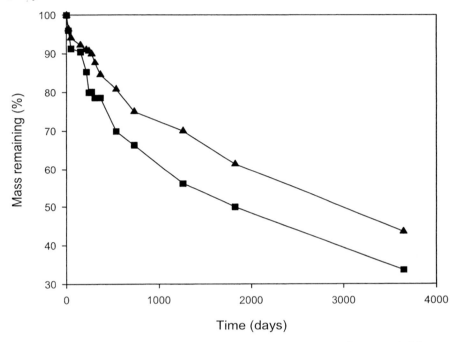

Fig. 9.4. Mass remaining in fine roots incubated in litterbags over a 10-year period in a sugar maple forest floor on Black Hawk Island (see Appendix III). (▲) sugar maple; (■) white pine (C. McClaugherty unpubl.)

Studies in boreal and temperate forests have also found that the chemical composition of fine roots influences their decay rates. For example, Camiré et al. (1991) noted that fine root decay rates appeared to be inversely related to initial N concentrations. They based this on their own findings with black alder and hybrid poplar roots and on comparison with results of Berg (1984) and McClaugherty et al. (1984).

9.3.3 Changes in chemical composition

Fine roots are chemically similar to foliar litter, so one would expect that their organic chemical composition would change during decay with a pattern similar to that for foliage litter. Although many studies have reported initial chemical composition of fine roots, few have followed the chemical composition during decay.

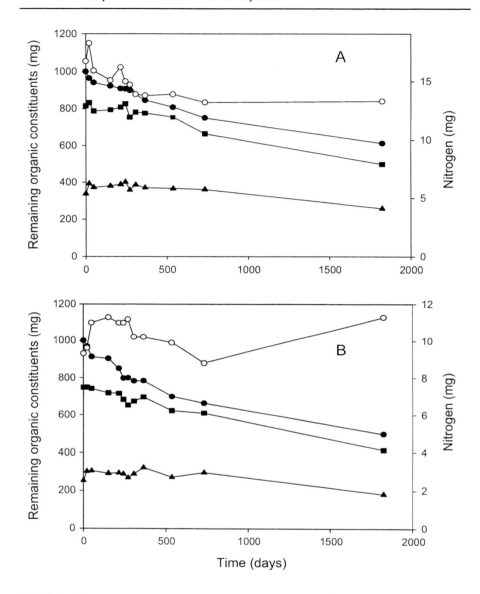

Fig. 9.5. Changes in amounts of some organic constituents and N over 5 years in fine roots incubated in litterbags on the forest floor of a sugar maple forest on Black Hawk Island, (see Appendix III). Organic constituents and nitrogen are expressed as mg g^{-1} of initial material. **A** Sugar maple fine roots. **B** White pine fine roots. (●) Total organic matter; (■) lignocellulose; (▲) lignin; (○) total N. (C. McClaugherty unpubl.)

The changes in chemical composition of fine roots during 5 years of decay in litterbags in a temperate sugar maple forest (Black Hawk Island, Appendix III) are shown in Fig. 9.5 (McClaugherty unpubl.). For white pine and sugar maple, most of the mass loss during the first year could be explained by the loss of soluble substances. Subsequent losses were largely due to a decline in polymer carbohydrates. Lignin declined rather slowly, beginning after the first year of decay.

The concentration of N in the root litters was initially high, 16.7 and 9.3 mg g^{-1} in sugar maple and white pine, respectively. Nitrogen contents varied during decay, but there was no clear pattern as is generally seen for foliar and wood litters (Fig. 9.5).

10 Models that describe litter decomposition

10.1 Introduction

Models serve a variety of functions for ecologists, but all ecological models are abstract representations of biological systems expressed in either mathematical or symbolic terms. Models may serve as hypotheses to be tested, or as tools to predict behavior of the ecosystem or one of its subsystems. In a concise review, Moorehead et al. (1996) distinguished three groups of decomposition models.

Some models are empirical and most of these are statistically based. For example, regression models relate parameters in a system. These models are useful for identifying or indicating the strength of hypothesized relationships, but cannot, by themselves, reveal causality. Empirical models are often useful for prediction, but they are, or at least should be, limited to the range of data from which they were developed. Extrapolation, however tempting, can be misleading.

Mechanistic models are another general class of models. They are often analytical in nature, using a system of equations to describe complex processes. Such models have proven very useful for gaining insight into ecosystem behavior, and for developing and testing general theories.

A third group of models are simulation models. They are created to simulate the behavior of a system, in a way that allows researchers to manipulate initial conditions, or other aspects of the model, to investigate potential outcomes. Simulation models may use a combination of mechanistic and empirical components to achieve their goal. Ecosystem simulation models often include a decomposition submodel.

In reality, models are often hybrids of these types. Empirical or data-based models may be based on a mechanistic understanding of processes. From a different perspective, theoretical (theory-based) models can be made more specific by validation with experimental data, and by using data to determine parameters. Simulation models often use such hybrids to enhance their ability to predict.

One of the challenges facing those who model decomposition is the large number of factors that influence the rates and patterns of litter decomposition. Thus, a single model or a relatively simple approach would not likely give a generally applicable description of the decomposition process. Factors that influence decomposition can be highly interactive, variable, and even hard to measure. These factors include microbial ingrowth, climate, variation in weather between years and different levels of nutrients and lignin. Considering the complexity of the decomposition process, we should expect a set or system of functions, each specific for different litter types and conditions. Such a set of functions in relation to litter type and ecosystem remains to be established. In this chapter we simply describe empirical functions that have been found to fit particular types of litters.

The concept of "kinetics" for litter decomposition is not used here in the same way as it is for enzymatic or chemical reactions, for which there are well-defined systems with e.g. zero order, first order or second order reactions. Rather, the mathematical descriptions of the litter decomposition process have used functions that simply fit to, and describe the process, as well as possible. These fits are sometimes related to short-term decomposition, covering just part of the process, often to less than 50% mass loss. The models describe the decomposition as regulated by the sum of different effects on the decomposition process. Effects that are often used in these models are litter chemical composition, soil or site richness and variation in climate over the period of incubation .

There are, of course, several mathematical functions that may be used to describe the litter mass-loss process (Howard and Howard 1974), but the models we present here have been the most commonly used. The most widely used mathematical model - the single exponential model (Jenny et al. 1949; Olson 1963) - is often used for very early stages of litter mass loss, but frequently unable to fit observations from later phases of decomposition. A slightly more complex model uses a double-exponential with two decay rate constants. These two decay constants can relate to two different components of litter, or to two different phases of decay (Lousier and Parkinson 1976). A further development of the idea that it is necessary to use differing rates for different substrate-quality compartments, is the triple model (Couteaux et al. 1998). This model divides litter into three substrate-quality components, each with different decay rates, and estimates separate rate constants for each. With the different rate constants that are produced from this model, it has been possible to estimate the potential decomposition rates for the very late stages.

A different approach is represented by the model based on an asymptotic function that estimates the decomposition rate as the derivative of the function at each point of the graph, and ultimately reaches an asymptote, which is normally at a level between 50 and 100% mass loss.

Investigators working from a mechanistic perspective have proposed general models that could also lead to an asymptotic function. Among the first was Carpenter (1981). His model was based on the idea that litter constituents could be placed on a continuum of decomposability. Furthermore the model allows that during the decay process, particular components could be transformed into substances of either higher or lower decomposability. Carpenter's model produced declining decay rates, and provided improved fits to the decay data for aquatic vegetation.

Bosatta and Ågren (1991) developed a different mechanistic model. Their model is based on the heterogeneity of the litter substrate, the efficiency of the microbial community, and the changing accessibility of the substrate to the decomposer community over time. Their model predicts different outcomes that depend on the nature of the microbial community and the composition of the decomposing litter. Under some conditions, their model predicts that litter will leave a finite residue with no further decomposition.

In this chapter we focus on empirical models, though as stated above, the best empirical models have a mechanistic foundation. In practice, different types of models may be used for the same data set, and the fit of theoretical to observed data will give different levels of statistical significance. However, the utility of a given model as a predictor is dependent not only on a statistical significance of the fit, but also on the causal relationships that are behind the specific model.

10.2 Two main kinds of empirical models

The models found in the literature may be divided into two categories. The first group of models comprises those that describe the decomposition of the whole litter, specifically whole litter mass loss. The organic matter that is being degraded is thus regarded as one "unified" material. Sometimes, the ash is subtracted, for example at high concentrations, and the model is applied onto the organic matter only (Faituri 2002). We may call them unified-substrate-quality models.

A second group of models was developed from litter that had been analyzed for organic-chemical components of different stabilities. For example, if two main groups of organic matter are degraded at very different rates, these degradation rates are estimated separately for each component. Similar models with three components have been developed (Lousier and Parkinson 1976; Couteaux et al. 1998). Mathematical formulae for examples of these models are given in Table 10.1.

Table 10.1. Some models used to describe the decomposition of litter

Formula	Comments	Characteristic	Reference
Unified-substrate quality			
$M_t = A + Br^t$	Asymptotic	leaves a residual	[a]
$L_t = m(1 - e^{-kt/m})$	Asymptotic	leaves a residual	[b]
$M_t = M_0 e^{-kt}$	Single exponential	leaves no residual	[c, d]
Two or three substrate-quality components			
$M_t = Ae^{-k_1 t} + Be^{-k_2 t}$	Double exponential	leaves no residual	[e,f]
$M_t = Ae^{-k_1 t} + Be^{-k_2 t} + Ce^{-k_3 t}$	Triple exponential	leaves no residual	[g]

[a]Howard and Howard (1974)
[b]Berg and Ekbohm (1991)
[c]Jenny et al. (1949)
[d]Olson (1963)
[e]Bunnel et al. (1977)
[f]Lousier and Parkinson (1976)
[g]Couteaux et al. (1998)

With regards to the litter types for which these models are applicable, we can distinguish two main classes: foliar and non-foliar litter. The above models have been applied primarily to foliar litter. The two major types of non-foliar litter are wood and roots. For wood, a critical characteristic is the extremely low nutrient concentrations, especially that of N (see Chaps. 4 and 9), that changes the decomposition pattern. We give attention to wood and root decomposition in Chapter 9. In the present chapter we intend to focus on models that describe decomposition of foliar litter.

10.3 Models used to describe decomposition of whole litter as a single or "unified" substrate

10.3.1 Single exponential

This model, first proposed by Jenny et al. (1949), and elaborated by Olson (1963), is an equation for first-order kinetics, the same as for radioactive decay. A basic condition for applying this equation is that the process runs at the same rate (constant fractional rate), irrespective of the amount of material left at any given point in time, and that one component is considered as active in the process.

The formula may be written (Wieder and Lang 1982):

$$M_t = M_0 e^{-kt} \tag{10.1}$$

and is often used in this form

$$ln(M_t / M_0) = -kt \tag{10.2}$$

In these and subsequent equations, M_0 is the initial mass, M_t is the mass at a certain time, t, and k is the decay rate constant. The single exponential model is often used for predictive purposes, based on the assumption that the decomposition rate is constant and that all material is decomposed. The "half time" and "mean residence time" of litter is also calculated, although the validity of this function in any specific case is open to question. Aber et al. (1990) suggested that this model works reasonably well for a variety of litters until only 20% of initial mass is remaining. Because of its relative simplicity and its reasonably good fit for early stage of decay, this model is widely used (Gholz et al. 2000).

10.3.2 Asymptotic model

We have described (Chaps. 2 and 6) that for several litter types, decomposition proceeds progressively more slowly, and may even approach zero, as decay progresses, which is most likely due to the retarded degradation of lignin. Howard and Howard (1974) found that the amounts remaining after decomposition of some litter types approached a minimum level. They found that the model which best described this process was an asymptotic non-linear model with three parameters: A, B, and r.

$$M_t = A + Br^t \qquad (10.3)$$

M_t is the percentage of remaining litter mass, t is time in days, A and B are variable parameters, and r is an expression for the decomposition rate. By definition, the sum of A and B should be equal to 100 %, resulting in only two free parameters. By making a slight parameterization ($m=B$ and $k=B$ $\ln r$) Berg and Ekbohm (1991) arrived at the following non-linear model that they found more feasible to use (Fig. 6.3):

$$L_t = m(1 - e^{-kt/m}) \qquad (10.4)$$

where L_t is the accumulated mass loss (in percent), t is time in days, k is the decomposition rate at the beginning of the decay, and m represents the asymptotic level that the accumulated mass loss will ultimately reach, normally not 100% and often considerably less. The k of this function is the derivative of the function at $t=0$ and should not be directly compared to rate constants estimated with other models.

The asymptote should not be regarded as an asymptote in a strict mathematical sense, but rather as a practical limit for decomposition. The asymptote can be related negatively to, among other factors, initial litter N concentrations and positively to litter Mn concentrations. These nutrients may regulate the microbial degradation of the litter's lignin. and N is an active participant in the formation of humic acids that may retard the decomposition (cf. Chap. 6).

10.4 Dominant factors that influence the unified-substrate models

Numerous factors, both internal and external to the litter, and related to the collection of data, can influence how well a unified-substrate model can be extended across litter types or ecosystems. For example, levels of lignin and macronutrients are not consistent within or across litter types. The quality of the collected data, and the duration and frequency of the measurements, can also influence the development of a model. Before applying a model, several questions must be answered. Given a model based on a particular litter type with a certain chemical composition, would the model be appropriate over a range of ecosystems? How much would a selected characteristic of the ecosystems, for example nutrient availability in the humus layers, influence the fit of the model? Further, how would climate influence the model?

To our knowledge there are no consistent answers to these questions. In addition, the specific microbial population of a site, having developed in relation to the local environment, is a critical factor. Thus, for a given litter type, the decomposition pattern may vary between different systems and climates. Generally, we expect all factors that influence decomposition rates to also have a potential influence on the general decomposition pattern.

Different models may be applied to the same data sets with varying degrees of fit. One point in choosing a model is how far the decomposition pattern will be described. The technique of measuring litter decomposition as mass loss may allow the decomposition to be followed until ca. 60-70% mass loss, or until the process has come to a halt, which for some litter types may take place even earlier (see Chap. 6). The present basis for our discussion is that decomposition should have either reached a point where no further change can be measured, or be followed to at least 60-70% mass loss.

10.4.1 Extent and quality of the data set

It is often the patterns in later stages of decay that are most difficult to describe with a model. As a rule, data sets that cover only small values of accumulated mass loss fit well to the single exponential model. Likewise, small data sets with a low number of measured values can almost always satisfy a single exponential equation. In contrast, data sets that are sufficient to test for asymptotic functions have a set of conditions on them. Thus, a data set with a low number of mass-loss values is not likely to give a significant limit value (asymptote). Since data often are collected over some years, annual and seasonal variation in weather will influence the mass-loss patterns. The best data sets tested so far have been those with a high number of samplings, ideally ten or more, and with some of the samples collected at exactly one-year intervals to minimize the effect of short-term weather variations. The quality of a data set is often determined by the number of replicates for each sampling relative to the inherent variability of the data. For example, in the case of litter bags, a low number of replicates normally results in a scatter among the average values. Further, the study should ideally follow the accumulated mass loss far enough so that the measured values are within 20% of the asymptotic value.

10.4.2 Substrate quality

Several litter types decompose more quickly when the litter is newly shed, but the rate decreases in later stages as the litter "ages". This process has been discussed above (Chaps. 2, 5, 6) and may be caused in part by changing concentrations of lignin, Mn and N, factors that may have a causal relationship to the retardation.

In some cases, the retardation of decomposition can be traced back to the initial concentrations of these components. An example shows a clear effect of different nutrient levels on mass-loss rates as described by a single exponential. Six sets (Table 10.2) of Scots pine needle litter all have highly significant fits to the single exponential model when all mass-loss data are used, consisting of 10-12 samplings and up to 70% mass loss. All six sets have similar k values, ranging from 0.26 to 0.32 year^{-1}. This indicates similar overall rates for decomposition, with no

Table 10.2. A comparison of k values estimated using the constant-fractional-rate model given in Eq. (10.2) (Olson 1963), for an early stage (m.l. <40%), for a late stage (40%<m.l. <70%), and for early and late stages of decomposition combined (m.l. <70%). R^2 values within parenthesis. Using the asymptotic model and mass loss values for both stages combined (Eq. 10.4), the initial rates and limit values were estimated (Table 10.1). All litter types compared are Scots pine needle litter with different nutrient levels. Please note that the magnitudes of the k values (B and C) and the initial rate (D) are not comparable. **A** Litter type and initial composition. **B** Single exponential model - all data combined. **C** Single exponential model - two phases. **D** Asymptotic model. (B. Berg unpubl.)

A. Litter type and composition		Initial conc. [mg g^{-1}]	
Designation	Litter / treatment	N	P
1	Brown / unfertilized	4.0	0.21
2	Brown / unfertilized	4.4	0.32
3	Brown / fertilized 40 kg N/year for 6 years	4.4	0.3
4	Brown / fertilized 80 kg N/year for 6 years	7.0	0.34
5	Brown / fertilized 120 kg N/year for 6 years	8.1	0.42
6	Green	15.1	1.31
B. Single exponential model – all data combined			
Designation	m.l <70%, n=10 – 12		
	k values [year^{-1}]	Intercept	
1	0.2949 (0.980)	0.0128	
2	0.3103 (0.936)	0.0829	
3	0.3019 (0.948)	0.0790	
4	0.3179 (0.959)	0.0806	
5	0.2964 (0.911)	0.1365	
6	0.2602 (0.960)	0.1382	
C. Single exponential model - two phases			
Designation	m.l.<40%, n=4-5	m.l.>40%, n=5-6	
	k values [year^{-1}]	k values [year^{-1}]	
1	0.2949 (0.976)	0.2303 (0.984)	
2	0.3989 (0.953)	0.2029 (0.936)	
3	0.3880 (0.966)	0.2059 (0.969)	
4	0.4073 (0.977)	0.2267 (0.989)	
5	0.4592 (0.980)	0.1723 (0.928)	
6	0.4709 (0.991)	0.2025 (0.980)	
D. Asymptotic model			
Designation	Initial rate [% day^{-1}]	Limit value [%]	
1	0.0768	93.2	
2	0.1087	78.2	
3	0.1055	77.4	
4	0.1112	78.0	
5	0.1299	72.2	
6	0.1360	68.0	

trend in the k values, even though there is a clear trend in litter chemical composition. For example, N increases from 4.0 to 15.1 mg g^{-1} and P from 0.21 to 1.31 mg g^{-1}. Further, there is no relationship between nutrient levels and k values. The approximation that is made when using all data, results in a consistently higher intercept as the linear regression adapts to the data sets (Table 10.2B). The increasing curvature with increasing nutrient levels is seen in Fig. 10.1.

Comparing these results to those obtained when the sets are split into an early phase with mass loss < 40%, and a later phase with mass loss >40% but < 70%, reveals greater variability among the six sets. The separation into phases made it possible to distinguish trends in the data, and showed that k values for high-nutrient litter with mass loss < 40% are almost double those of low-nutrient litter. For brown litter, the initial rate (k, Eq. 10.2) is related to the initial concentration of nutrients such as P ($R^2=0.950;n=5$). When calculating the rate for later stages only (mass loss >40%), the rates are much lower than in the early phase (Table 10.2) and no trend is seen. Thus, splitting the single exponential into two phases may help us to resolve the process, and we may uncover large differences in rate between early and late phase of decomposition Table 10.2). This example also indicates the limitations of the single exponential model.

Figure 10.1 shows an example with three sets of Scots pine needle litter with different levels of nutrients, and we have compared the pattern of the first order kinetics graphs to the N levels. Figure 10.1A shows a needle litter with low initial concentration of N (4 mg g^{-1}), one line shows the observed values for $ln\ (M_t/M_0)$ for this litter, and the other the constant fractional rate extrapolated. The two lines clearly cover each other and there is no trend towards a retardation of the decomposition rate. Figure 10.1B shows a litter with an initial N concentration twice as high. We can see that the measured values start deviating after ca. 700 days of incubation (at ca. 58% mass loss). For a set of needle litter with almost four times as high an initial N concentration (Fig 10.1C) the deviation begins even earlier, namely after ca. 400 days of incubation (ca. 38% litter mass loss). For the two more N-rich litter there is a clear deviation from the single exponential model.

There is thus a clear deviation from the constant-fractional-rate model with increasing nutrient levels. The reasons for this could be a higher initial rate in the early stages, namely with increasing concentrations of N and P, which would be the effect of a limiting nutrient. There could also be a rate-retarding effect of raised N levels on decomposition in the later stages, an effect that is related to the degradation of lignin. It is likely that both effects are active, resulting in higher initial rates and lower ones in late stages. Thus, for a particular litter type, the higher the N level the larger the deviation from a single exponential. The decomposition for all the litter discussed above may be described by an asymptotic function (Table 10.2).

So, for what litter types would the constant fractional rate model be valid? The model fits well to the most nutrient-poor litter (Table 10.2D, Fig 10.1). Although the fit is based on empirical findings; we may still speculate that the nutrient levels may play a role as was suggested. As seen above, the mass-loss graphs for these litter types discussed above may be described by an asymptotic function that clearly indicates that the rate decreases to become close to zero. The fits with asymptotic models to the six sets of Scots pine needle litter are all highly significant, indicating that the model can describe the decreasing rate of the litter decomposition, and incorporates the extremely slow decomposition in the very late stages of decomposition (Table 10.2.D). The example above dealt with needle litter from

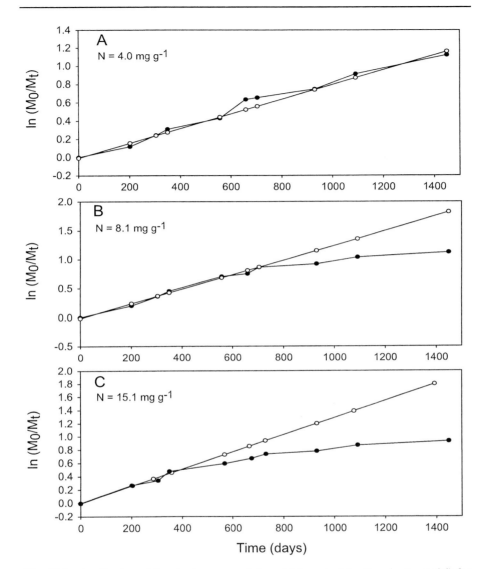

Fig. 10.1. Application of the single exponential model (constant fractional rate model) for Scots pine needle litter of three different N levels. On each plot, one line is estimated using all available data, specifically litter mass loss values from 0 to ca. 66-70% for litter of different initial N concentrations (●), and the other is an extrapolation of the fractional rate as estimated for the early stages (○). See Table 10.2. **A** For natural litter from a nutrient-poor plot, the rate was rather constant as seen in the comparison of the two graphs. **B** At a higher nutrient level (N concentration 8.1 mg g^{-1}), deviation from a constant rate starts at ca. 58% litter mass loss. **C** At an even higher nutrient level (N concentration of 15.1 mg g^{-1}), a deviation starts even earlier, at ca. 38% litter mass loss (B. Berg unpubl.)

Table 10.3. A single exponential model was fitted to mass loss values for seven litter types incubated under identical conditions but having different initial concentrations of N. All litter types had 12 to 13 samplings and the linear relationships were all highly significant (except for alder, $p<0.01$). The maximum incubation time was 1393 or 1448 days. (In part from Berg and Ekbohm 1991)

Litter type	Init N	Init P	Obs.	Single exp	Asymptotoic function	
	[mg g^{-1}]	[mg g^{-1}]	[%]	k [yr^{-1}]	k$_{init}$ [y^{-d}]	Lim val [%]
Scots pine br	4.8	0.33	67.4	0.299	0.093	89.0
Lodgepole pine br	3.9	0.34	63.1	0.281	0.075	100.0
Silver birch br	7.7	1.05	59.9	0.164	0.187	56.9
Silver birch gr	17.4	1.80	63.0	0.182	0.206	54.3
Scots pine gr	15.1	1.31	65.9	0.259	0.136	68.0
Lodgepole pine gr	10.5	0.82	66.0	0.256	0.086	81.5
Grey alder gr	30.7	1.37	55.5	0.12	0.336	50.6

br brown, *gr* green, *Obs..* maximum observed mass loss, *Single exp.* single exponential, *yr* year, *d* day, *Lim val* limit value

Scots pine only. When making a similar comparison using different litter types (Table 10.3), a similar result emerged, namely that the deviation from the simple model increased with increasing nutrient levels. We may also see that the rate constants mainly decrease the more rich in N and P the litter is. In comparison, the asymptotic function (Eq. 10.4) gives initial rates that mainly fit to the initial nutrient levels and the limit value for decomposition (Berg and Ekbohm 1991).

This asymptotic model has been tested against approximately 150 data sets from foliar litter decomposition studies (Berg and Johansson 1998). It has given significant asymptotic limit values for practically all these data sets. For some decomposition studies using nutrient poor litter species, the model has not given significant limit values, although sets of mass loss values from identically designed experiments using more nutrient-rich litter, and run in parallel, have given significant fits. Thus, Berg et al. (200X) found that brown lodgepole pine needle litter, poor in N and relatively rich in Mn, did not give a significant limit value. That study was carried out in two contrasting forest stands: one temperate with wet and warm climate, and rich in N and other nutrients; the second in a boreal system, with low temperatures, rather dry and in combination with a nutrient-poor soil.

In earlier studies, Berg and Ekbohm (1991, 1993) found that the same litter type (lodgepole pine needle litter) gave a limit value at or near 100% decomposition. There is a clear trend in this material indicating that the lower the litter N concentration, the higher the limit value (Berg 2000b; Berg et al. 200X), and we cannot exclude that one of the limitations of the model is related to the levels of N and possibly Mn in the litter. Lodgepole pine needle litter from trees grown in Scandinavia is also rich in Mn (Chap. 4), which we have hypothesized to promote the degradation of lignin and litter (Chap. 6).

The models that so far best describe the long-term decomposition of the main foliar litter types are asymptotic ones (Table 10.2, Fig. 6.3). This does not mean that they have any generality, only that they fit very well to those foliar litter types investigated so far.

As mentioned above, a condition for using the single exponential model is that the litter is decomposing as a unified substrate. Thus, the decomposition of a litter for which the decomposition process is suppressed by, for example, slow lignin degradation, or accelerated by a large initial concentration of solubles (Figs. 2.2 and 6.1) would not be well described by a single exponential approach. However, that may be a matter of how much of the decay process is being described. In the example of Scots pine needle litter, decomposition in the early phase was not suppressed by the increasing lignin concentration, and a single exponential describes this early phase well. Figure 7.6 shows that the decomposition of the same litter, incubated under different climatic conditions, is hampered to different extents. For a litter incubated close to the Arctic Circle at an annual average temp of ca. 0-1°C, approximately 600 mm annual precipitation and an AET of 387 mm, raised lignin concentrations had very little effect (Berg et al. 1993b). In contrast, at a more southern site the effect of lignin was very pronounced. Although an analysis of such a climatic influence on decomposition kinetics has not yet been made, we predict that the single exponential model may be applied to a greater fraction of the decomposition process in colder climates.

Since lignin and N in combination have proven to be suppressing agents, we speculate that for a foliar litter poor in N, it would be possible to apply a single exponential to a higher mass loss than for a more N rich litter. Taylor et al. (1989) successfully used the single exponential function for a set of litter types with initial N concentrations ranging from 5.2 to 13.1 mg g^{-1}, and accumulated mass losses from 22.3 to 56.1%.

10.5 Models based on two or three substrate-quality fractions

10.5.1 The double exponential

The double exponential model is a development of the single exponential and is based on the assumption that the litter substrate has two main substrate-quality components with different decomposition rates. Its construction is simply an addition of two factors each defining a part of the litter substrate.

$$M_t = Ae^{-k_1 t} + Be^{-k_2 t} \qquad (10.5)$$

In which t is time and k_1 and k_2 are rate constants for quickly and slowly decomposing fractions of the litter. The amount of each fraction is given by A and B, respectively. The model appeared for the first time in the literature from the International Biological Program (IBP) Tundra Biome (Bunell et al. 1977), and in a paper by Lousier and Parkinson (1976) describing the decomposition of aspen leaf litter.

10.5.2 The triple exponential

The triple exponential model is a development of the double exponential and is based on the assumption that the litter substrate has three main components with different decomposition rates instead of two. Its construction is simply an addition of three exponential factors for decay.

$$M_t = Ae^{-k_1 t} + Be^{-k_2 t} + Ce^{-k_3 t} \tag{10.6}$$

in which t is time and k_1, k_2, and k_3 are rate constants for quickly, slowly, and extremely slowly decomposing fractions of the litter. A, B, and C, give the amount of each fraction, respectively.

Couteaux et al. (1998) used this model to describe the decomposition of Scots pine needle litter and estimated the rates of three different fractions. They divided the substrate into "labile", "metastable", and "recalcitrant" fractions, and obtained k-values in the range from $10^{-1}\%$ day^{-1} to $10^{-4}\%$ day^{-1}. They also applied this model to respiration of humus samples. A drawback of this model is that it assumes invariable k-values, an assumption that is questionable when applied to the later stages of decay.

11 Decomposition and ecosystem function

11.1 Introductory comments

The microbial decomposition of plant litter is a basic process in the functioning of ecosystems, not only for the general release of nutrients to plants, but also for the buildup of a stable humus and the accompanying storage of nutrients.

Nutrients are released from decomposing plant litter, either through leaching from newly shed litter, or because of decomposition processes. Leaching, and the pattern of release due to decomposition, is dependent on litter type, its surroundings and the kind of nutrient. Thus, nutrient release is closely tied to ecosystem structure and function. We can distinguish two main pathways for nutrients: one is release and the other is storage in a stable form in the humus.

Humus accumulates as the stand grows and ages (Ovington 1959; Forrest and Ovington 1970; Bormann and deBell 1981; Schiffman and Johnson 1989). In these studies the accumulation followed a nearly linear increase with stand age. A model formulated to describe the linear accumulation of humus with time may be general, but lacks causality.

Humus accumulation rates may be estimated by a summation of the recalcitrant or resistant part of the decomposing litter. The resistant fraction is based on the concept of a limit value (Chap. 6) that measures the fraction of litter that decomposes extremely slowly. The level of the limit value is determined by causal factors, such as the concentrations of lignin, N and Mn in litter.

There is very good support for the observation that a long-term net accumulation of humus takes place, even over millennia (Jenny 1980; Wardle et al. 1997). Such an accumulation can be predicted using the limit value approach to humus accumulation in undisturbed systems (Berg et al. 2001).

Neither C nor N were present in the original mineral soil that existed before plants evolved, but entered the ecosystem from the atmosphere. Both C and N are macronutrients, and fulfill very different functions. Carbon makes up the skeleton of the macromolecules that create a storage matrix for N and other nutrients. Nitrogen is a major nutrient that must be stored in the ecosystem, to supply plants with an even flow of mineral N. Loss of N from the ecosystem would cause the vegetation on a given plot to move to an earlier succesional stage due to N limitation, and the ecosystem may need N_2 fixing organisms to restore N to its prior level.

The mechanism by which the ecosystem stores N depends on the structure of the litter produced by a given plant, the degree to which that litter decomposes and the transformations it undergoes during decomposition. It would be reasonable to expect that each plant species would produce litter and litter remains that would store nutrients in concentrations high enough to allow the species to survive. A collection of plants making up an ecosystem may, for the purpose of survival, need a feedback mechanism based on nutrient availability.

Nutrient elements can be added to an ecosystem through abiotic weathering, aerosol input from outside the system, or fertilization. Ecosystems have the capacity to store at least some of these added nutrients. This chapter aims to evaluate what a storage mechanism for humus means in terms of the ecosystem. It also aims to describe N dynamics, in terms of the relative amounts of N being released *vs.* those being stored. The possible effects of climate change on such a storage mechanism are also discussed. Of course, this latter section is an extrapolation of existing, and partially empirical, data and must be regarded as a prediction that needs to be validated.

11.2. Humus is accumulating in undisturbed forest ecosystems.

11.2.1. How far can humus accumulate?

The increasing amounts of humus measured over a stand age by Ovington (1959) were relatively low, reaching about 4 kg m^{-2} after 55 years. As already mentioned, Jenny (1980) and Paul (1984) suggested that such an accumulation would continue for millennia in the absence of disturbance, which would result in the buildup of considerable amounts of humus.

That humus can accumulate over considerable periods, if disturbances such as fire and forest management are excluded, is now an established fact, and amounts as high as 109 kg m^{-2} in a temperate forest system have been found after an unknown duration of accumulation (South Italy, Berg et al. 1993a). Forest floor masses of 49 kg m^{-2} have been reported after 3000 years of accumulation in a boreal forest, resulting in humus layers of up to 1.5 m thickness (Wardle et al. 1997, Table 11.1). Such values may give us a perspective on the age of humus layers that have accumulated over shorter periods. In boreal Scandinavia, layers deeper than 20 cm are common, and more than 9% of about 15,000 sampled plots (Swedish Forest Inventory) had humus layers deeper than 40 cm (Table 11.2), and almost 8% had a depth of more than 50 cm, suggesting an accumulation for periods up to a millennium. The concept of "Tangelhumus," found in the Alps and other central European mountains, with a humus layer up to 1 meter thick as part of the definition, further supports a real long-term accretion (Rehfüss 1990; Anonymous 1996).

A clear relationship is seen between the frequency of forest fires and humus accumulation (Wardle et al. 1997). With such heavy layers of humus being formed when undisturbed, it is reasonable to speculate that fire is the normal, and potentially the dominant, agent of humus layer reduction. Recent measurements of extremely low humus respiration rates (Couteaux et al. 1998), indicate that a tremely low humus respiration rates (Couteaux et al. 1998), indicate that a balance between humus formation from litter input and decomposition is unlikely within millennia (Chap. 6).

Table 11.1. Observed and estimated amount of humus of known age in north Swedish forests at ca. 66°N (Wardle et al. 1997) in a Scots pine forest at the SWECON site Jädraås at 60°49′N (Berg et al. 2001). Given are also measured amounts of N and P

	Northern Sweden - Islands			Jädraås
	<0.1 ha	0.1-1.0 ha	>1.0 ha	
Age [years]	2984	2081	1106	120 [a]
Forest floor mass [kg m^{-2}]	49.08	34.62	14.33	1.54
Increment [kg m^{-2} year^{-1}]	0.0164	0.0166	0.0130	0.0128[a]
Est. litter fall[b] [kg m^{-2}]	0.08-0.14	0.08-0.14	0.08-0.14	--
Modeled litter fall 112 years [kg m^{-2}] [a]	--	--	--	151.55
Est. limit value [c] [%]	87.8	87.6	90.4	92.1
Av. limit value[d] [%]	83.0	83.0	83.0	89.0 [e]
Est. forest floor mass[f] [kg m^{-2}]	41.2-72.1	28.7-47.2	15.2-26.7	1.67
Missing fraction[f] (needle litter basis) [%]	16	17	6	--
Excess fraction [f] (total litter basis) [%]	46	37.6	86.3	8.4
N storage [g m^{-2}] [i]	761	460	163	15 [h]
P storage [g] [g m^{-2}]	39.1 [g]	28.7 [g]	9.2 [g]	0.72 [h]

[a]It was estimated that at this site it would take 8 years before the litter could be considered to be humus and part of the F/H layer (Berg et al. 1995b).
[b]Litter fall for the three Hornavan groups was estimated using available Scandinavian data for pine and spruce forests between 59°N (north of the line Oslo-Stockholm-Helsinki) and 67°N (Berg et al. 1999a, 2000). The lower value gives needle litter fall and the higher total litter fall.
[c]This limit value estimated as 1- increment/needle litter fall.
[d]Estimated from existing limit values (n=18) for Scots pine and Norway spruce litter at sites in northern Sweden (Berg and Johansson 1998).
[e]Limit value for needle litter decomposition at site Jädraås (Berg et al. 1995b).
[f]Berg et al. (2001).
[g]D. Wardle (pers. comm.).
[h]Calculated from Berg et al. (200X)
[i]Wardle et al. 1995

The comparison of humus accumulation rates among four stands suggests that there was no real difference in rates between the young stand, 120 years old, and the older ones (1106 to 2984 years, Table11.1), which supports the idea that an undisturbed system is not likely to reach a steady-state in less than several millennia. In addition, we have not found any real support for the concept of a steady state in humus accumulation. We would suggest that the end of the growth of a humus layer, or its destruction, is more likely to be dependent on disturbances or catastrophic events such as fire.

The accumulation of humus should be regarded in the context of the complete ecosystem. The effects on humus accumulation of different soil systems under growing trees will be discussed. A forest stand that has been clear-cut represents a rapidly changing ecosystem. Its soil system alters rapidly due to increasing levels of available nutrients, increasing moisture and a probable change in the microbial community. The following discussion of humus accumulation will therefore focus on soils under continuously growing trees.

Table 11.2. In the Swedish Forest Inventory 14,234 forest plots were investigated in the period from 1984 to 1987 with measurements that included humus depth. The frequencies below do not include mires or mountain plots. Swedish Natl Survey of Forest Soils and Vegetation, Dept of Forest Soils, SLU

Humus thickness	Frequency	Frequency
[cm]	number of plots	[% of total]
0-10	8838	61
10-20	3008	21
20-30	750	5.2
30-40	305	2.1
40-50	214	1.5
50-60	202	1.4
>60	917	6.3

11.2.2 A mechanism for humus accumulation under undisturbed conditions

Litter chemical composition and limit values

When Howard and Howard (1974) and Berg and Ekbohm (1991) showed that litter decomposition appeared to come to a halt, they estimated limit values that were significantly different between the litter species they investigated (Chap. 6). Their work was based on the assumption that the sum of decomposition processes resulting in litter mass-loss would continue to slow down without dramatic breaks in that pattern. In Chapter 6 we discussed the limit value concept for decomposition of foliar litter, and the effects of litter N and Mn concentrations on the extent of decomposition, as well as the amount of recalcitrant remains.

When limit values for foliar litter decomposing in natural systems were regressed against concentrations of nutrients and of lignin, it was seen that litter N concentrations gave a highly significant and negative relationship (Fig. 6.10, Table 6.6). The fact that in this large data set, the relationship to N concentration was significant indicates a general effect of N over a good number of species in deciduous and coniferous ecosystems in boreal and temperate forests.

Although limit values for litter mass loss have been estimated for a variety of litters using asymptotic functions, such limit values may not necessarily indicate that the remaining organic matter is completely undegradeable by biological agents (see below). The residual organic matter could very well consist of a stabilized fraction that decomposes very slowly, or a fraction that does not decompose in a given environment, but would undergo further decomposition after a change in that environment. The discovery of an apparent final mass-loss value should not be considered trivial, however, especially if the limit value could be related to litter properties, such as lignin concentration, nutrient status or climatic factors. When Berg et al. (1996a) compared C storage in the humus layer in paired stands of Norway spruce and Scots pine, the measured humus buildup could be related to different limit values for the two species, which in turn were related to different N levels (see section below). In a review, Cole et al. (1995) compared the organic matter buildup under red alder (high litter N level) and Douglas-fir (low litter N level). When Berg et al. (2001) evaluated their data, using the limit value approach they found a much higher store of organic matter under the former than could be explained by simply comparing litter fall. Thus, the existing data support the hypothesis of a mechanism based on the limit value concept to estimate organic matter buildup. Still, a set of experiments is needed to confirm the single steps in the mechanism.

Table 11.3. Concentrations of nutrients in ash-free humus in a boreal nutrient-poor Scots pine stand (Jädraås) and a richer, temperate silver fir stand (Monte Taburno). The values are based on ash-free organic matter. (Berg et al. 200X)

Concentration of nutrient										
[mg g^{-1}]							[µg g^{-1}]			
N	P	K	Ca	Mg	Mn	Fe	Zn	Cd	Cu	Pb
Silver fir, Monte Taburno										
38.2	2.84	17.7	20.0	4.76	0.7	26.5	0.1	0.8	62.	9.7
Scots pine, Jädraås										
11.8	0.47	10.9	3.23	0.98	0.2	9.40	0.0	0.6	9.2	8.9

Forest humus systems have different levels of nutrients. Large differences exist between systems (Table 11.3), for example a Scots pine system has a humus N level of 11.8 mg g^{-1} while a silver fir system has a humus N level of 38.2 mg g^{-1}, and generally higher levels of other nutrients. The soil microorganism community will have adapted to these different nutrient levels (cf. Chap. 3). In a very N-rich system a higher percentage of the lignin-degrading organisms would not be sensitive to N or less sensitive as compared to N-poor systems, which may mean that the limit values could be higher. In such a system, the concentration of N may have little or no effect on the degradation of lignin and on the limit values, but levels of heavy metals may have a greater impact (Berg et al. 200X).

Nitrogen, Mn, and heavy metals may be system-specific indicators

The fact that significant linear relationships exist between limit values and initial litter concentrations of both N and Mn, and that there are causal explanations for these relationships, offer support for a strong rate-regulating mechanism. The role of the initial concentrations of these nutrients is based on empirical relationships, and while they can be regarded as an index, they are not necessarily general. Although we have obtained a general relationship over several ecosystems, such relationships need to be proven and confirmed for other types of ecosystems.

Additional stabilizing factors

Litter at the limit value is biologically stable in undisturbed systems. With decomposition rates as low as 0.0001 to 0.00001% per day, values of the same magnitude as those measured from humus of the same stand, we can regard the litter as stable humus (e.g. mor humus). As such, it can build humus layers over millennia. There are, however, additional stabilizing mechanisms that could contribute to creating a humus layer. One such mechanism would be an increasing anaerobicity in the thickening humus layer. The thickness of the layer itself would cause difficulties for diffusion of oxygen, and the ability of humus to hold water would cause anaerobic pockets to develop. Lack of oxygen could prevent complete metabolism by aerobic microorganisms, and instead of releasing carbon dioxide and water as final products, organic acids would be produced. An increase in the anaerobic microflora would produce similar products that could either inhibit or resist further degradation. Organic acids alone (e.g. acetic acid and benzoic acid) are well known inhibitors of microbial activity and, combined with a lower pH, could reduce the rate of litter decomposition, thus enhancing the rate of humus accumulation.

Litter components change with stand development

Not all litter components behave the same as foliar litter in terms of humus formation (Chap. 9), and the types of plant material contributing to litter fall changes with stand age (Fig. 11.1).

At the Scots pine case-study site, litter fall was observed for 7 years in three Scots pine stands, which were 18, 55 and 120 years of age at the start of the study (Flower-Ellis 1985). There was an increase in total litter fall in all three stands due to both increased foliar litter fall and the addition of more woody components with age (Berg et al. 1993d). Part of the increase in litter fall with stand age may be attributed to an increase in tree biomass and increased foliar litter fall. However, as trees reach physiological maturity, cones develop, and part of the increase in litter fall is due to the addition of cones. Bark and twigs normally start falling later, in this case at an age of about 22-23 years, thus increasing the proportion of woody components with stand age.

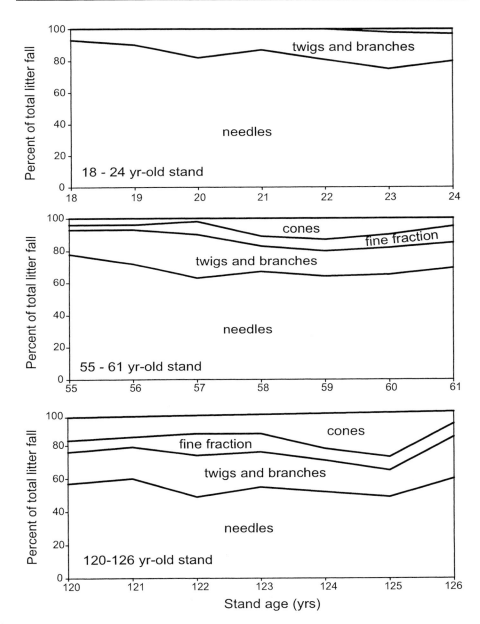

Fig. 11.1. Relative proportions of needles, fine litter, twig and branch litter and cones during the period 1973/1974 to 1979/1989 in stands starting at 18, 55 and 120 years of age. (Berg et al. 1993d; J. Flower-Ellis 1985)

What litter components can form humus?

The scientific literature has not yet thoroughly addressed the question of how the various litter components are transformed into humus. There appears to be an understanding that humus is formed from litter as if plant litter was chemically and structurally homogeneous within a species. According to the existing definitions of humus (e.g. Bal 1973), all the components of litter are equally capable of forming humus.

The type of litter influences the type of humus that is formed. For example, mor, moder or mull humus types in boreal systems have been partially investigated and explained empirically (e.g. Lundmark and Johansson 1986). Those discussions focused on the species from which different types of humus are formed, and not on the causal reasons for humus formation as based on long-term stable compounds. This observed difference between plant species as regards formation of humus types may in part have its origin in the difference in chemical composition, something that directly influences the microbial communities and soil animals, and thus the decomposition process (Chaps. 6 and 9).

This background may allow us to draw the conclusion that a large part of the foliar litter components form humus according to the schedule suggested in Chapter 6. In contrast, the information concerning the decomposition of the more nutrient-poor (especially N-poor) components is such that we judge that their contribution to humus is more uncertain. Relatively few studies are published about the decomposition kinetics of such material, but Harmon et al. (1986) have conducted extensive work on wood decay. Their studies have included branches, stumps, and stems and the results (M. Harmon, pers. comm.) may be summarized thus: the decomposition of N-poor woody material is very dependent on which microorganisms, mainly fungi, are first to colonize the wood. If white-rot fungi dominate as invaders, the decomposition goes rapidly and progresses to completion, leaving minimal remains. If, on the other hand, brown-rot fungi dominate in the attack, the decomposition will not be complete and a residue will persist. Brown-rot fungi, through their sheer dominance, can prevent ingrowth of the more efficient white-rot organisms. Following a brown-rot attack, a fragmented humus-like material is left that has a very low turnover rate, and can be found in the humus layer for a long time (Fig. 11.2).

Foliar litter types with different N concentrations have stable remains in proportion to their N levels (Chap. 6). The extremely low N levels in different kinds of woody litter result in dramatically different decomposition processes as compared to foliage. Woody material which is attacked by white-rot may be decomposed rather quickly leaving little recalcitrant residue, while woody material attacked by brown-rot, may turn into more stable organic matter (Chap. 9).

An observation on the decomposition of woody material by Johnson and Todd (1998) illustrates the potential contribution of wood within a temperate mixed oak forest. They state that dead stems constitute a very common and visible component on the ground in older coniferous systems in northwestern USA. Thus, it is often assumed that this woody debris has an important role in the functioning of

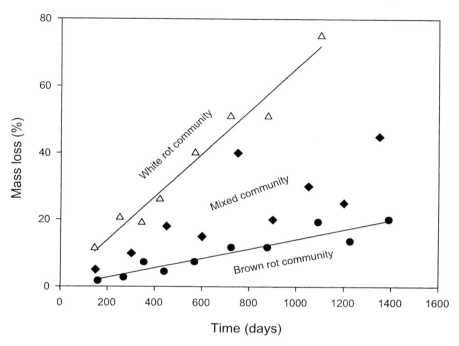

Fig. 11.2. A hypothetical decomposition pattern for woody litter material depending on whether it is being degraded by white-rots (Δ), brown-rots (●), or both (♦)

of the ecosystem, including a habitat for animals, microorganisms, and new plants. However, they conclude that dead stems constitute a relatively unimportant component in the mixed oak forests in southeastern USA. Much of the larger diameter woody residue that was left on the ground, decomposed by more than 80% after only 15 years. They claimed that this residue had no effect on the C-contents of the soil.

We can conclude that in the above study the magnitude of the contribution of woody components to the formation of humus was possibly low, at least for the humus layer, since the leaching was not studied or discussed. We may further conclude that the quantitative contribution of wood to humus among boreal systems is unclear, and that the decomposition patterns that govern the importance of the contributions are unknown. Although over-simplified, we can speculate that systems dominated by white-rot fungi should result in a soil system in which a complete decomposition of woody materials dominates. Likewise, in systems in which brown-rot organisms dominate a proportionally larger part of the humus would be based on woody materials. This reasoning could probably also be applied to woody root litter, although whether roots form significant amounts of humus is uncertain (Chap. 9). However, the dominance of one group of fungi or the other in different systems is not well known.

Woody components are decomposed towards humus in a pattern different from that of the more nutrient-rich foliar litter components. The attack on wood may be random, or ruled by environmental factors that either give one or the other kind of lignin-degrading organisms an advantage, on the basis of e.g. location, season, state of decay or wood species. The general nutrient status of the stand can of course affect which organisms would dominate. When remains of partly decomposed woody material (often a brown powder) are seen in the humus layer, this should not be interpreted as the only kind of decomposition, and does not mean that woody materials generally form humus. Further, we do not know what proportion of the total inflow of woody material remains in the humus as partly decomposed material. For mature boreal stands, about 30% of the litter fall is not foliar but consists largely of woody matter (Mälkönen 1974). Thus, in a system in which white-rot fungi dominate, we could expect that less would be left of such material, and its contribution to the humus layer smaller compared to system in which brown-rot dominates. On the other hand, in boreal coniferous forests, brown-rots generally dominate (T. Nilsson, pers. comm.), which would mean that a higher fraction of woody material would stay longer as recalcitrant material. However, the longevity of this recalcitrant material of woody origin is unknown, and we cannot tell if it forms humus in the same sense as foliar litter.

In the case of fine (<1mm diameter) roots, the picture is not very clear. We have, for example, not found any direct studies on decomposition of fine roots. In addition, our own studies, as well as personal communications with other scientists, indicated that traditional decomposition experiments do not seem to give a true picture of mycorrhizal fine root decomposition. For example, dead fine roots of pine that were incubated in litter bags inside the humus layer have shown a strong resistance to decomposition. In contrast, studies have shown that the fungal component of the mycorrhizae of pine fine roots may be involved in the decomposition of humus. This indicates that the fungal component of mycorrhizae may have an important role as a decomposer organism (section below). In addition, large and rapid fluctuations in fine-root biomass indicate that their decomposition may follow different patterns than that followed by foliar litter (Chap. 9; McClaugherty et al. 1982).

The main contribution to humus from the fungal biomass surrounding the fine roots may be from dead and recalcitrant fungal mycelia. This would represent a very different mechanism for addition of material to humus.

Foliar litter forms humus, and we can distinguish a connection between its chemical composition, and the potential for humus formation. The picture is less clear for the more nutrient-poor litter components, such as woody litter and woody waste. We are unable to quantify their contribution to humus, probably because of a lack of knowledge about the ecology of the lignin-decomposing organisms.

11.2.3. An accumulation mechanism can be validated

The question of humus stability is connected to the ecosystem, its organisms and changes taking place in the system. In the discussion below we focus on humus stability in undisturbed forest ecosystems, under continually growing forest. Using the growing forest as a condition on the ecosystem, we have used long-term accumulation of humus as a measure of its stability.

Three approaches have been used to test the validity of the limit value concept as a tool for estimating humus accumulation (Berg et al. 2001): (1) a direct comparison between measurements of accumulated humus and storage estimated by the limit value concept; (2) the effect of species, in this case foliar litter with different N concentrations, and mass of litter fall, were compared to humus storage; and (3) the estimated N concentrations at the limit values were compared with the humus N concentrations in the same stands.

To verify the limit value model for long-term C accumulation, conditions must be set on the test system. Such conditions are not always easy to fulfill. Both the correct quantitative information and the site history should be available (Berg et al. 2001). The accumulation of humus may thus be estimated theoretically using the knowledge of limit values and the magnitude of litter fall. Because the N level of the litter largely determines the limit value and it may be used to make an estimate of humus accumulation (Fig. 11.3; Berg 2000b; Berg et al. 1996b).

Direct humus measurements

Site data and history for a 120-year-old stand. A budget for humus was created for a Scots pine forest, and validated with information from the same pine forest, for which there existed well-documented background data and site history (Berg et al. 1995b). For this forest site, there are extremely good data for litter fall, litter decomposition and amounts of soil organic matter on the ground. An important fact is that a violent fire took place in the mid-1800s that burnt off the existing organic layer. The existing humus layer has been built up on the resulting ash layer from the litter fall of the existing stand.

A litter-fall model estimated litter fall for 120 years. Litter fall was measured for 7 and 10 years in each of two pine stands, which were 18 and 120 years old, respectively at the start of the study. The stands were growing on the same soil, and had the same climate and hydrology (Flower-Ellis 1985). The combined litter fall measurements covered a period of 17 years. This wide sample allowed the possibility to adapt two litter-fall models (below) for a mature stand. Root litter was not considered, as the pine roots had been observed mainly in the mineral soil in this stand (H. Persson, pers. comm.), and only lingonberry rhizomes and heather roots, which form a small fraction of the total root biomass, were found in the humus layer.

Litter fall usually increases with stand age until a maximum biomass is reached. Data from a chronological study (above) can lead to a mathematical description of litter fall over a stand age. We consider two alternative models for litter fall (Fig. 11.4) as a function of stand age, using the above data. The first, and simplest, model assumes a linear increase in litter fall until canopy closure, after which it remains constant. For example, annual litter fall would increase in equal steps of 16.2 kg ha^{-1} each year from year 1, reaching a constant input of 1,620 kg ha^{-1} yr^{-1}

in year 100. This was the measured 10-year average amount of litter fall in the stand (Berg et al. 1995b).

The second model assumes a logistic increase in litter production. The logistic model can be stated as:

$$\frac{dF}{dt} = gL(\frac{F_{max} - F}{F_{max}})$$ (11.1)

It can be solved to give:

$$F_t = \frac{F_{max} F_0}{F_0 + (F_{max} - F_0)e^{-gt}}$$ (11.2)

where: F_0 = annual litter fall at time $(t) = 0$, F_t = annual litter fall at time = t, F_{max} = maximum ("steady-state") annual litter fall, g = constant intrinsic for rate of increase in litter fall with stand age. The unit for litter fall was kg ha^{-1}.

Using serial approximations to achieve the best fit to data from both the 18-year-old and 120-year-old stands, the following parameters were derived: F_{max} = 1620, g = 0.37.

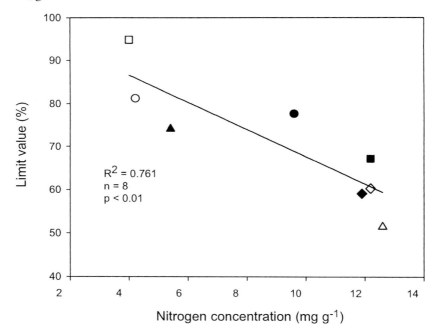

Fig. 11.3. The limit value for decomposition and thus the magnitude of the recalcitrant (humus-forming) part of the litter is related to initial N concentration in litter. Average values for eight litter species: (○) Scots pine - brown needles, (■) Scots pine - green needles, (□) lodgepole pine, (▲) Norway spruce, (△) silver fir, (●) silver birch, (♦) European beech, (◊) oak spp. (Berg (2000b)

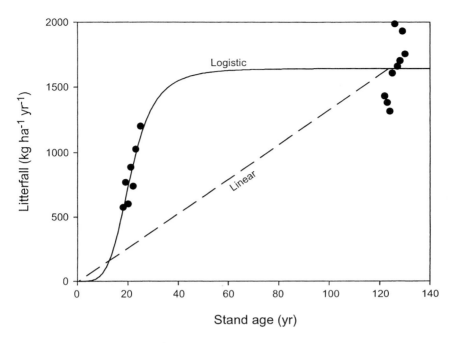

Fig. 11.4. Two simplified models for litter fall with stand age developed using data from a chronosequence of Scots pine stands. One assumes a linear increase in litter fall with age, up to 100 years, followed by a constant litter fall. The other uses a logistic, nonlinear model fitted to the data for the 18-25 and the 120-130 year old stands in a chronosequence (SWECON site, Jädraås, Appendix III). (Berg et al. 1995b)

A budget was set up. Staaf and Berg (1977) determined that the amount of SOM was 1.54 kg m^{-2} in the combined A_{01}-A_{02} horizon. This value, based on ash-free matter, did not include distinguishable litter remains. The needle litter fall from Scots pine completely dominated the litter inflow. Based on eight measurments of the limit value this needle litter left an estimated residual of about 11% (cf. Chap. 2; Berg et al. 1995b). Using that as a basis, the litter inflow for each of 112 years was estimated (the litter formed in the last 8 years had not yet formed a stable humus). Addition over 112 years gave an estimate of the accumulated organic substance of 1.67 kg m^{-2}. This theoretical result differed by only 8% from the observed level of 1.54 kg m^{-2} (Staaf and Berg 1977; Table 11.1). Because the chemical composition of the litter can be important in determining the limit value, it is important to note that the foliar litter formed from the pines was chemically similar over different years (Table 4.3, Berg et al. 1993b). Furthermore, the main part of the litter from the shrubs on the site had a chemical composition (lignin and N) that was close to that of the pines (Berg and Staaf 1981).

Three budgets covering millennia. Berg et al. (2001) set up further budgets by using data from Wardle et al. (1997). The average age of three groups of humus layers from three separate forest stands had been determined to be 2984,

2081, and 1106 years, using ^{14}C-analysis of ash from the latest forest fire. Wardle et al. (1997) determined the amount of ash-free humus to be 49.0, 34.6, and 14.3 kg m^{-2} respectively (Table 11.1).

A simple regression model for needle litter fall from the completely dominant species (Scots pine and Norway spruce), based on measurements in boreal stands (northern Europe, between 52°N and 67°N), was used to estimate litter fall for all three groups of stands. With the use of the average limit value for needle litter of Scots pine and Norway spruce in northern Sweden (n=18), the fraction remaining was calculated, and the magnitude of the annual accumulation was estimated. The annual accumulation was summed over the different periods. The accumulation thus estimated for foliar litter was 41.2, 28.7, and 15.2 kg m^{-2}. The first two values were 16 and 17% lower than the observed values, whereas for the 1106-year-old humus the estimate was 6% too high. When using total litter fall, the estimates were generally too high with 46, 37.6, and 86.3% respectively for the 2984-, 2081-, and 1106-year-old stands (Table 11.1). It is possible that the method used to estimate litter-fall is susceptible to error, and this may in part explain the magnitude of the deviation. Only data for litter fall were used, while the root-litter component was not considered.

Different species - different N levels - predicted differences in paired stands of Scots pine / Norway spruce and Douglas-fir / red alder

Eight paired stands of Scots pine and Norway spruce were analysed in one experiment. The paired stands were of the same age, they were growing on the same soil, and under the same climates. In each stand, litter fall was measured, the chemical composition of litter determined and the amount of C in the humus layer on top of the mineral soil measured (Berg et al. 1996a).

Berg et al. (1996a) found that the needle litter of Norway spruce had, on average, higher N levels than Scots pine, with 5.16 mg g^{-1} in Norway spruce litter and 4.08 mg g^{-1} for Scots pine. This difference was statistically significant over all stands located at eight climatically different sites.

Limit values for spruce needle litter were lower than for pine. The average limit values for Norway spruce and Scots pine in these stands were 77 and 81.2%, respectively. This would correspond to a potential storage of 23 and 18.8% of foliar litter fall for spruce and pine, respectively.

The quantity of litter fall is one factor that influences humus buildup. An evaluation of the litter-fall data suggests that the pine plots have a significantly higher needle litter fall. However, there was a larger accumulation of humus in the spruce than in the pine stands. When determining the amount of C that has actually accumulated in the humus of the investigated stands, Berg et al. (1996a) found that on average 13.4% more humus was stored in spruce compared to pine stands. The spruce forests thus had a higher accumulation of C in spite of the higher litter fall in the pine forests. A theoretical estimate, based on an equal litter fall, suggested that humus accumulation should be 26% higher in the spruce stands as compared to the accumulation in pine. It is reasonable to suggest that the higher litter fall in the pine forests had reduced the difference between the theoretical and observed values.

Humus accumulation is lower in a stand poor in N (Douglas-fir) than in an N-rich one (red alder) on the same soil. A recalculation of a study (Cole et al. 1995) of two stands of equal age and on the same soil, one of which was Douglas-fir (N level 5.4 mg g^{-1} in the foliar litter), and the other red alder (the N level of the foliar litter was 23 mg g^{-1}), showed that the latter had a clearly higher amount of soil C than the former. Berg et al. (2001) estimated the limit value for decomposition by using the N concentrations of the foliar litter, and constructed a model for the build-up of the soil organic-matter layer. From the model estimate, it was evident that the higher accumulation of humus/soil C in the alder stand could not be explained only by a higher litter fall. In table 11.4 the difference in estimated amounts of SOM (using the limit-value concept) was 7625 g m^{-2} and the difference in the measured amount 7300 g m^{-2}, the latter figure being the sum of SOM in the O horizon, wood, and SOM in the mineral soil. A good part of the explanation may be found in the different limit values that were estimated from the different N levels (Table 11.4).

Table 11.4. Amounts of litter fall and its N concentration as well as accumulation of ash-free SOM in soil under 50-year-old red alder and Douglas-fir stands. The data, from Cole et al. (1995) and Johnson and Lindberg (1992), have been converted to g m^{-2} of organic matter. (Berg et al. 2001)

Fraction	Red alder	Douglas-fir	Difference
Nitrogen in litter fall			
N conc. in foliar litter [mg g^{-1}]	18.8	7.4	
Amount of litter fall [g m^{-2}]			
Leaf litter	312	105	
Non-leaf litter	136	73	
Understory litter fall	126	46.6	
Estimated accumulated litter fall over 50 years	25980	15105	
Limit-values and estimates			
Estimated limit values [%][a]	56	74.8	
Estimated accumulation over 50 years [g m^{-2}]	11431	3806	7625
Measured amounts of SOM [g m^{-2}]			
O horizon	7960	2350	5610
Wood	1890	1320	570
Min soil 0-7 cm	6040	6180	
Min soil 7-15 cm	3640	4040	
Min soil 15-30 cm	6880	5580	
Min soil 30-45 cm	6400	6040	
Total in min soil 0-45	22960	21840	1120

[a]Using the equation, limit value = -1.6474 N + 86.95 (Berg et al. 2001)

Comparison between estimated N concentration at the limit value for decomposition and that measured for humus in the same stand

The litter N concentration increases linearly over the course of litter decomposition, and is highly correlated to accumulated litter mass loss (cf. Chap. 5; Fig. 11.5). By estimating and using limit values, and the linear relationship between N concentration in litter and its accumulated mass loss, the N concentration at the limit values can be estimated (Fig. 11.5, Berg et al. 1999b). The estimated N concentration at the limit value could thus be compared to that measured in the ash-free mor humus in the same stand. Nitrogen could thus be used as an internal marker to confirm the limit value.

For a larger number of litter studies (local needle litter) the limit value for decomposition was calculated. By extrapolation of the linear relationship between accumulated mass loss and the increasing litter N concentration, Berg et al. (1999b) estimated the N concentrations (Fig. 11.5). These estimated N concentrations were compared to those of the humus of the A_{01} and A_{02} layers in the same stands. Forty-eight stands with eight tree species were used.

In a comparison between the estimated and measured N levels for humus, they found that the estimated values for the N concentration in humus were 6.8% lower than those observed (Fig. 11.5B). For example, if the estimated humus-N concentration were 1.00%, then the measured concentration would be 1.07%. This difference has the same magnitude as the methodological error for the N analysis. Berg et al. (1999b) concluded that the litter remains at the limit value had a relatively long-term stability, and indicated that there was a possibility of estimating humus N buildup (below).

11.2.4 Can different ecosystems accumulate humus at different rates?

From the above it follows that humus accumulation is dependent on both the magnitude of the litter fall and the chemical composition of the foliar litter (above, Chaps. 2 and 6). The magnitude of the litter fall and the distribution of litter components give the quantity, while the chemical composition determines the substrate quality, and thus the magnitude of the limit value and the stable fraction. The term "stable fraction" must be used with some reservation. It has been demonstrated that humus that is stable in its ecosystem, and with the predominant microbial community, can be suddenly decomposed, probably due to a change in the mycorrhizal system (below; Sect. 11.3.2). The term "stable humus" should rather be considered as "potentially stable humus".

Different species

The general relationship between limit values and initial N levels in the foliar litter (Fig. 6.10) may be elaborated. The linear regression observed is based on 21 spe-

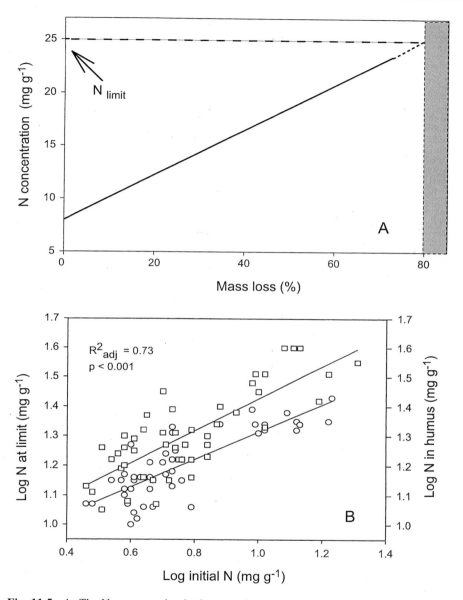

Fig. 11.5. **A** The N concentration in decomposing litter increases linearly to the accumulated mass loss. The N concentration at the limit value (N_{limit}) is estimated through linear extrapolation (Berg et al. 1999b). **B** A comparison of the estimated N concentration at the limit value (○) and the actually measured one in the ash-free humus layers (□) in the corresponding forest stands. Data from 48 decomposition studies with local foliar litter were used (Berg et al. 1999b)

cies, but 80 of the values originate from just 8 species (Table 11.5). Using these eight litter types and comparing the mean values for N concentrations and limit values give us clear borders for these litter types, and result in a highly significant correlation (R^2=0.760; Fig. 11.3) between average N concentrations and average limit values. Although the litter types have been collected over rather large regions, the differences among them in the concentration of componentssuch as N, remain consistent (cf. Chap. 4). Thus, comparing the species, we see that boreal lodgepole pine would, on average, leave a small fraction of about 5.1% of the shed needle litter, while temperate silver fir would leave 48.8%, giving a tenfold variation in remaining fractions (potential humus; Table 11.5). This indicates a clear importance of tree species in determining the amount of humus accumulation.

Berg et al. (2001) defined potential storage efficiency, using the remaining fraction of the decomposing litter and the litter fall. They called it the litter-to-humus (LH) factor and it is calculated as:

$$LH = (100\text{-limit value})/100 \qquad (11.3)$$

This would give:

$$\text{Litter fall x LH} = \text{accumulated humus} \qquad (11.4)$$

Table 11.5. Average limit values for decomposition and N concentrations for eight litter types. The potential capacity to accumulate humus (Pot. humus) is given by (100-limit value)/100. The potential capacity to store N is given and defined at the limit value (Pot N stor. = N concentration at limit value × fraction remaining). In part from Berg (1998b) and Berg and Dise (200Xa)

Litter	Initial N	Limit value	Pot. humus	N limit	Pot. N stor.
	[mg g^{-1}]	[%]	[fraction]	[mg g^{-1}]	[mg g^{-1}]
Lodgepole pine	4.00	94.91	0.051	13.60	0.69
Scots pine (brown)	4.19	81.3	0.187	12.76	2.39
Silver birch	9.55	77.7	0.223	22.71	5.06
Norway spruce	5.44	74.07	0.269	14.46	3.75
Scots pine (green)	12.18	67.2	0.328	--	--
Pyrenean oak	12.2	60.3	0.397	--	--
European beech	11.9	59.12	0.409	24.05	9.84
Silver fir	12.85	51.5	0.488	21.93	10.64

Pot. humus potential fraction becoming humus (the LH factor), *N limit* N concentration at the limit value, *Pot. N stor.* potential amount of N stored in humus at the limit value per gram of initial litter

Assuming a similar magnitude for foliar litter fall, the LH factor would give the relative amounts of litter remains being stored, and there would be an extreme factor of 10 between lodgepole pine and silver fir in terms of potential capacity to store humus (Table 11.5). When comparing the two pine species, there is a factor of 3-4 between lodgepole pine and Scots pine, and one of 1.4 between Scots pine and Norway spruce.

Foliar litter fall may vary among systems. Even within a species there are general relationships both with the nutrient availability of a system and the climate. Richer systems and those in warmer, wetter climates have higher litter fall (Berg and Meentemeyer 2001).

There may be differences in annual litter fall between species even if both the soil and climate are the same. Berg et al. (1996a, 2001) compared needle fall between eight paired stands of Scots pine and Norway spruce, and found a 7-8% higher annual litter fall in the Scots pine stands. Since both the amount of litter fall and the LH factor are variables in Eq. (11.4), a high LH factor may have a higher influence than the amount of litter fall. Thus, in spite of the higher annual litter fall in Scots pine stands described above, the Norway spruce stands formed more humus than the Scots pines (cf. Table 11.5).

Differing litter chemical composition

Two chemical components, N and lignin, have an overall role in the retardation of litter decomposition (Berg and Matzner 1997). When plotting available data for initial concentrations of N and lignin for foliar litter, Berg (2000b) found that the litter types formed distinct groups (Fig. 11.6). The litter species formed these groups in spite of the fact that the sampling occurred over a large geographical region the magnitude of Scandinavia. For Scots pine, the sampling included stands of very different nutrient status. Thus, Scots pine needle litter formed a homogeneous group that did not overlap with the lodgepole pine or Norway spruce groups (Fig. 11.6).

In this comparison, Scots pine needle litter has low concentrations of both N and lignin, whereas lodgepole pine litter had low N and high lignin concentrations. Norway spruce needles formed a group that had higher N concentrations than those of the two pine species, and lignin concentrations that were in between. Silver birch leaves had lignin concentrations similar to those of the Norway spruce needles and generally higher N concentrations. The leaves of European beech, which were collected from the whole of Western Europe, formed an extreme. In a substudy, needle litters from Norway spruce, lodgepole pine and Scots pine were collected annuall in adjacent stands on the same soil. The variation within each group was of the same magnitude as within the larger regions (B. Berg unpubl.). The effects of such differences are reflected in limit values among species (Fig. 11.3 and above).

We can conclude that plant species contributing to the chemical composition of litter fall is a major factor affecting humus formation, at least in natural, unpolluted systems. The limited within-species variation in concentrations of the two main components, lignin and N, over a large region, supports this. Still, the variation in Mn and Ca concentrations may influence the level of the limit value and the accumulation of humus.

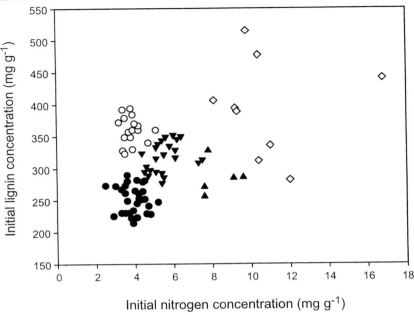

Fig. 11.6. Concentrations of lignin and N in newly shed foliar litter of Scots pine (●), lodgepole pine (○), Norway spruce (▼), silver birch (▲) and European beech (◊). All available data. Redrawn from Berg (2000b)

11.3 How stable is humus?

11.3.1 Do limit values indicate a complete stop in litter decomposition?

Although it is possible to estimate significant limit values for litter decomposition, we cannot conclude that such limit values necessarily indicate completely recalcitrant remains in the humus-near litter. Couteaux et al. (1998) applied both a three-factorial model (Chap. 10), and a limit-value function, to direct measurements of decomposing Scots pine needle litter, as well as to the humus formed in the same stand. They measured k-values for decomposition of a stable fraction, close to the limit value, of 0.0001 to 0.00001% per day, which corresponds to a rate of about 1% per 30 to 300 years. That study included an analysis of stable, meta-stable, and labile components (Table 11.6). Of those three fractions, the stable fraction comprised about 80% of the total organic material and could be considered as rate limiting. The estimated k-values may thus indicate a fraction that is well stabilized, and thus decomposing at a very low rate (cf. below). The fact that allophanic (see Glossary) humus exists shows that an "eternal" storage is possible. Although allophanic organic material may be regarded as an extreme case, the level of stabilizing components (e.g. aluminum and iron) necessary to stop the decomposition process is not known (Paul 1984).

Table 11.6. Fractions of different size and stability in far-decomposed Scots pine needle litter and humus from the Scots pine stand of incubation. Data from Couteaux et al. (1998). The k-value given by Couteaux et al. (1998) has been recalculated to the corresponding values for mass loss in % year^{-1} for the sake of comparison to values for litter decomposition

Fraction	Size [%]	k [% day^{-1}]	Mass loss [% year^{-1}]
Far-decomposed litter			
Labile	5	0.124	30
Meta-stable	15	0.087	3-6
Stable	80	10^{-4} to 10^{-5}	0.03-0.003
Humus			
Labile	5	0.124	30
Meta-stable	15	0.087	3-6
Stable	80	10^{-4} to 10^{-5}	0.03-0.003

11.3.2. Four classes of humus turnover

We can distinguish four main classes of turnover and major influences on the decomposition of humus. First, there is the class of humus decomposition that occurs in completely undisturbed humus. There are two classes associated with elevated microbial activity: one being caused by a strongly activated mycorrhizae; the other due to mechanical disturbances that are caused by soil manipulation and drainage leading to radically higher decomposition rates. Analogous disturbances that occur when sampling humus probably explain the high rates of CO_2 efflux sometimes reported for humus (e.g. Persson et al. 2000). Finally, high rates of humus turnover have been observed in humus subjected to very high N inflows.

Humus decomposition - undisturbed systems

Respiration (release of carbon dioxide) is commonly used to measure humus decomposition and normally includes a disturbance of the sample, which can lead to an increased respiratory activity. A major problem with describing soil respiration is the difficulty of separating root respiration, faunal respiration and microbial respiration in intact systems (Hanson et al. 2000; Högberg et al. 2001).

There are few direct gravimetric measurements on humus decomposition, and we have not found any study that has given long-term data for humus decomposition *in situ* and in undisturbed samples. A requirement for a correct measurement is that it includes ash-free humus rather than whole humus layers, and that the system should not have been disturbed. An additional problem is that only part of the material classified as humus in the humus layer is stabilized (Townsend et al. 1995, 1997; Olsson et al. 1996; Couteaux et al. 1998), while a good part of the substance classified as humus is considerably more easily degradable than the stabilized fraction. For example, Staaf and Berg (1977) found that, of the substance that they classified as humus in Scots pine stands, about 15% was water soluble. In the humus of a different Scots pine stand, Couteaux et al. (1998) found a labile pool of about 5%, a metastable of pool of ca. 13-15% and a recalcitrant one of ca. 80-85% (Table 11.6).

Following a clear cut of a Scots pine forest, Berg and Staaf (1983) measured the disappearance of ash-free humus layers. A reinterpretation of their data, moving from a linear relationship to a more curvilinear one, indicates that nearly all of the 27% of the organic material that disappeared during 4 years, was probably lost in the first 2 years. This could mean that when remaining dead root material and soluble matter was decomposed, no measurable decomposition took part in the last 2 years.

Olsson et al. (1996) found mass losses ranging from 0 to 7% for pine forest humus, and 17 to 22% for spruce forest humus over a period of 15-16 years. If labile and meta-stable fractions were decomposed first, and were the main component of the mass loss recorded, this should mean that the Scots pine humus was stabilized to a similar degree as that found by Couteaux et al. (1998).

Decomposition rates measured in the laboratory (disturbed samples). There have been numerous respiration measurements (O_2 and CO_2) carried out on humus. However, almost any type of sampling involves some level of disturbance of the humus system and its microorganisms. When the humus is sorted, and roots or other inclusions are removed, the disturbance will be even larger. A disturbance of this kind normally results in greatly increased activity lasting for some days to a week. Because of this, samples are pre-incubated for up to 2 weeks to ensure they are "stabilized", before the measurements are carried out. This does not mean that they are so stable that they can be used to quantify humus decomposition and give results that accurately model those occurring in undisturbed soil conditions.

In an investigation of respiration studies, the authors used 20 randomly selected reports and recalculated the CO_2 release rates as percentage annual mass losses, to allow a comparison to be made among a large set of independent measurements. The resulting respiration values gave rates of between 5 and 30% C loss per year (e.g. Persson et al. 2000). The lower values are of the same magnitude as the measured values for decomposition of pine and spruce needles in late decomposition stages. Given that this sampled humus had been collected after an accumulation of tens or even hundreds of years, such high rates cannot accurately represent the respiration rate of the humus in undisturbed systems.

In their approach using respiration, Couteaux et al. (1998) investigated both humus and partially decomposed litter. During their study about 20% of the mass was respired, at rates that differed strongly between the labile and the stable pools.

The ratio between rates for the labile and recalcitrant pools was ca. 10,000:1, and the relative sizes of the pools were 5 and 80% of the total mass. These results suggest that many studies overestimate humus respiration as it would occur in an undisturbed system. In addition, in many of these experiments, it is likely that only the properties of a smaller, labile fraction of the humus sample will dominate the CO_2 release.

Comparing the orders of magnitude of the above estimates with those of Couteaux et al. (1998), we see that the higher rates (ca. 27% loss per year, Berg and Staaf 1983) coincide with those calculated by them for the labile fraction (ca. 30% loss per year). The lower rates correspond with the rates of the meta-stable fraction (ca. 3-6% per year). For the stable fraction (Couteaux et al. 1998), a rate of ca. 0.0001 to 0.00001% per year or ca. 1% in 30-300 years was found (Table 11.6). These calculations by Couteaux et al. (1998) deserve attention because they represent a new approach for experiments based on respiration measurements.

Disturbances of decomposition rates - some specific cases - mycorrhizae in natural systems.

In undisturbed humus systems, there appear to be mechanisms that can change the composition of the microflora in ways that enhance its ability to degrade the otherwise stable humus. Hintikka and Näykki (1967) gave a good description of the mycorrhizal basidiomycete *Hednellum ferrugineum* and its effects on the humus layer. The development of strong outbursts of soil respiratory activity, followed by a strong decrease in amounts of humus in the A_{01}-A_{02} layer was described. The effect was observed patch-wise on boreal dry, sandy, nutrient-poor sediment and till soils with a development of thick mycelial mats, and could be related to plant growth (see below). This is likely to be a powerful mechanism for humus decomposition that can transform potentially stable humus into a gray mycelial powder in a short time.

Thus, the same kind of humus in Scots pine forests that Couteaux et al. (1998) found to have an extremely low decomposition rate, could disappear almost completely in a very short time, leaving only a gray powder. In their pioneering study, Hintikka and Näykki (1967) supported their investigation with field measurements over humus that was almost completely degraded and compared respiration rates. An investigation into the frequency of this phenomenon of localized increases in decomposition, showed 13-14 such patches of about 0.5 to 2.0 m^2 in 1 ha of a very nutrient-poor Scots pine forest (B. Berg unpubl.).

In another system, Griffiths et al. (1990) studied the effects of the ectomycorrhizal fungus *Hysterangium setchelli* on respiration in humus under Douglas-fir and identified patches with very high respiratory activity. There appears to be a consensus among authors that the decomposition of humus material by ectomycorrhizal fungi provides an important source of nutrients for trees. This effect has also been observed for other mycorrhizal fungi (Unestam 1991). The current explanation is that when trees are subject to nutrient stress, their mycorrhizae are "activated" and switch their function from mutualists to decomposers. This means that nutrients are released in response to the trees' nutrient stress. The contrast between this activation and the more normal, low turnover leading to a fast accumulation is dramatic. However, the phenomenon appears sufficiently uncommon that it does not severely affect the kind of estimates of humus accumulation made by Berg et al. (1995b).

Disturbances of decomposition rates: fire, soil manipulation, and drainage

Fire is well recognized as a renewal agent for ecosystems, a general fact that does not need any further presentation, so we will focus on the effect of fire on humus layers. After mapping the wild-fire frequency over large parts of northernmost Sweden for more than 1000 years, Zachrisson (1977) found, for example, that in pine stands there was a fire every 50 to 60 years. Each fire removes at least part of the humus layer, which results in the generally thin humus layers, observed in pine forests in northern Fennoscandia. Normally, only parts of the humus layer would disappear in each fire, but in each fire, there was a mineralization of nutrients. This release of nutrients from the ashes may have a stimulating influence on the decomposition of the remaining humus layer lying underneath. Irrespective of nutrient effects, repeated fires have an influence on the accumulation and storage of humus. Thus, stand history is the dominant factor for determining humus accumulation rate (Wardle et al. 1997).

Disturbances of the soil system that increase decomposition activity have been noted in different contexts, for example in different kinds of site preparation, or simply the mixing of a soil sample. Johansson (1987) observed faster decomposition of Scots pine needle litter incubated under plowed-up humus and mineral soil in scarified plots, than compared to control litter. In an investigation of limit values, Berg and Johansson (1998) found that decomposition also went further after scarification of a plot's humus, and that significantly higher limit values were obtained for pine needles, increasing from 64% on nonscarified plots, to 97% for litter on scarified plots. Salonius (1983) made a careful investigation of material from L-, F- and H-layers, and found an increased respiration rate when these materials were mixed with mineral soil from the B horizon. Unfortunately, the mechanisms underlying the increased decomposition rate following disturbance are not known in detail.

Of the more large-scale observations, Delcourt and Harris (1980) compared the effects of the large American cultivation in the 18th and the 19th centuries with current practices, revealing what happens when the agricultural use of the soil is reduced. They concluded that in the past, cultivation caused disturbance of the soil and resulted in a large-scale release of C. Today that situation is reversed as the reduction in area under cultivation has turned the ground into a C sink.

Soil drainage increases the turnover rate of soil organic matter, as a result of both the direct effect of local disturbance, and the effect of a lowered water table. To explain this we may return to the conceptual model that illustrates the strong influence of lignin and lignin-like transformation products on decomposition of organic matter. These compounds cannot be degraded completely under anaerobic (oxygen-free) conditions, because the dominant decomposer organisms are fungi and oxygen-demanding bacteria. Ditching causes a drop in the water table, allowing oxygen to penetrate to deeper soil layers. This causes an increase in the degradation of the larger polymer aromatic compounds.

Disturbances of decomposition rates: systems with high N-loads

Observations have started to appear indicating a decomposition mechanism for humus that appears to be initiated by high acid or high N-loads of the soil system (Guggenberger 1994).   Such an effect was also mentioned by Nömmik and Vahtras (1982) in their review on ammonium fixation to organic matter in laboratory experiments. They referred then to the formation of soluble substances as an effect of the fixation.

Very high N-loads appear to promote a disintegration of humus, partly because of increased microbial activity (Fog 1988). Fog expressed the hypothesis that a higher concentration of N in litter resulted in an increased production of soluble organic matter (DOM or DOC). His ideas were based on the observation that lignin-degrading soft-rot fungi (Chap. 3) need, or at least tolerate, high N-levels in their surroundings. Therefore, in an environment rich in N, soft-rot fungi, to a certain extent, can replace white-rot organisms. Their degradation of lignin produces incompletely degraded lignin that can react with organic N compounds, which leads to water-soluble products that would then precipitate in the mineral soil. Fog's (1988) conclusion was that high N concentrations increase the formation of water-soluble, decomposition resistant compounds, but decrease the amount of humus that is formed. Ulrich (1981) described a similar process, and called it a "disintegration of humus". Other scientists, including David et al. (1989), have reported higher concentrations of soluble organic matter with increasing acidity. Guggenberger (1994) concluded that the mobilization of DOC is not ruled exclusively by a low pH. On the contrary, he concludes that high inflows of total N suppress the complete lignin degradation carried out by white-rot organisms, but increase the general microbial activity. He supports the conclusion proposed by Fog (1988) that the more N-tolerant soft-rot fungi produce partial degradation products, such as N-containing compounds, that are more water-soluble. Guggenberger also proposes that a generally higher microbial activity increases the production of microbial metabolites. In more recent studies of litter with moderate N levels, $^{13}CO_2$ release from litter was compared to leaching of ^{13}C from the same litter. The litter with higher N gave a higher leaching of C to the mineral soil (Flessa et al. 2000). It is possible that these studies indicate a general mechanism related to the N level of the litters.

To the above observations, we may connect a comparison of amounts of humus in mineral soil under Douglas-fir and red alder. In the paper by Cole et al. (1995) described earlier, it was shown that a higher amount of C compounds had been leached into the mineral soil from the humus layer in the N-rich alder stand. This single study cannot on its own support a general conclusion that an N supply in a naturally richer environment is part of a mechanism for formation of DOC that later precipitates in the mineral soil. Nevertheless, this observation also fits Fog's (1988) hypothesis, and the results of the ^{13}C laboratory study described above.

We may combine some of these observations to extend the hypothesis. Both Nömmik and Vahtras (1982) and Flessa et al. (2000) describe chemical transformations in humus that may produce water-soluble substances. The theory proposed by Fog (1988), concerning microbial components, may hold more generally, if modified. We may speculate that in humus exposed to high N-loads over a long period, a change takes place in the mycoflora (cf. Eriksson et al. 1990; Hatakka 2001). Lignin-degrading fungi that are not sensitive to, and may even be stimulated by, elevated N levels dominate the resulting fungal community. This would result in an acceleration of degradation processes, and would be dependent on the level of N deposition. This would also fit the observation by Guggenberger (1994).

11.3.3 Possible effects of increased temperature on humus decomposition- an artifact?

A general rule in ecology is Liebig's law of the minimum; a factor limits a physiological process only as long as no other factor overrules it (von Liebig 1847). Over the course of the transformation of litter to humus, there is at least one shift in such limiting factors.

The negative linear relationship between lignin levels and the decomposition rates of litter, change with climate, as indexed with AET (Berg et al. 1993e; Johansson et al. 1995). When decomposing litter approaches humus, the effect of climate decreases, and the effect of lignin levels increase to such an extent that the effect of climate was not measurable in experimental systems using litter-bags to model decomposition of an undisturbed organic substrate (Fig. 2.8B). The comparison was based on data from a 2,500-km-long climatic transect ranging from the Arctic Circle to northern Germany, with a variation in annual average temperature from 0.5 to 8 °C, and in AET from 357 to 559 mm.

Strömgren (2001) obtained a result that allowed similar conclusions by using an experimental system designed to measure the effect of soil temperature on soil respiration. After an initial period of 4 years during which soil temperature was kept 5 °C higher during the growth season than in the control, the soil-microbial system adapted to the artificial climate change, and no change in the CO_2 release from the ground was noted.

Rustad and Fernandez (1998) studied decomposition of red maple and red spruce foliar litter in heated (+ 5 °C) and control plots in a spruce-fir forest in Maine, USA. Red maple leaves lost mass more rapidly in the heated plot during the first 6 months, but no treatment effects on mass loss could be measured after 30 months. Red spruce litter mass loss was unaffected by heating during the first 18 months, but was greater in the heated plots (62%) than in the control plots (52%) after 30 months. In a similar soil-warming study in a northern hardwood forest in New York, USA, McHale et al. (1998) concluded that warming would have a short-term effect on CO_2 release until labile C was consumed. This result supports the observations made (Berg et al. 1993e; Johansson et al. 1995) for litter in humus-near decomposition stages, which suggest that factors other than climate dominate the decomposition rate of humus.

Support for this is also given by Berg and Matzner (1997), who reinterpreted data from Bringmark and Bringmark (1991) indicating a significant negative relationship between the N level in humus samples from a climate transect across Sweden, and their respiration rates. In a study over a climatic transect with coniferous forest stands, ranging from ca. 64°N in Sweden to ca. 40°N in Spain, Dalias et al. (2001) found that a standard plant material that had been incubated at a higher temperature, was more recalcitrant to C mineralization than the same material incubated at a lower temperature. Such an experiment indicates that some humus-stabilizing processes proceeded at a higher rate in the warmer climate, which lends support to the observations given above (cf. Sect. 7.4.2).

11.4 Storage of nutrients in humus

11.4.1 What amounts of nutrients can be stored in accumulating humus?

Different litter types have different concentrations of nutrients and heavy metals. Nutrients such as K are normally released at a high rate, with only a small proportion of the initial supply remaining in the litter (Laskowski et al. 1995). Several heavy metals (e.g. Pb) are, in practice, not released at all, and we could expect that decomposition is increasingly suppressed as the concentration of the heavy metal increases. There are, however, few studies in the literature concerning the effects of different nutrients and heavy metals at natural concentrations. We may have the best level of knowledge about N and as a result, we will focus on the effect of N.

Nitrogen

The dynamics of N show clear patterns that are easy to study and possible to relate to environmental factors. For example, it is well known that the concentration of total N increases during the decomposition of litter towards humus (cf. Chap. 5). The amount that is stored in a recalcitrant form can be estimated (Berg et al. 1999b).

When litter decomposes, its N concentration generally increases. Irrespective of the net dynamics, the concentration of N increases with litter mass loss, and this relationship is linear (Chap. 5). In systems that have not been artificially enriched in N, through either deposition or fertilization, the slope is repeatable and correlated to the initial N concentration (Berg et al. 1997). The N concentration will continue to increase as long as the litter is decomposing, and will only stop increasing when the limit value is reached (Berg et al. 1999b, Fig. 11.5). The linear relationship for the increase in N concentration relative to mass loss can be extrapolated to the mass loss value corresponding to the limit value, and thus give the N concentration that the litter will have when it reaches its stable phase. Using this data, it is also possible to calculate the amount of N that has been released from the litter, and the amount that will be stored in the stabilized litter (or humus). Using data from field studies, Berg et al. (1999b) calculated such values for a set of litter species. The potential capacity of different litter species to store N was defined by Berg and Dise (200Xa) as the amount of N remaining in stabilized litter or humus that was derived from 1 g of fresh litter. The capacity of the six different litter types studied to store N was related to their initial N concentrations. The richer the litter was in N, the more is stored (Fig. 11.7).

A more general validity of the N-storing mechanism may be expected over the boreal zone as judged from the data presented in Fig. 11.5B. This data, gathered from the whole of Scandinavia, represents a wide spectrum of N levels. As expected, litter with an increased N level formed N-rich humus, and the concentrations for N at the limit value and in humus were always higher than those in fresh litter (Berg et al. 1999b).

With knowledge of litter fall in a given stand, and the chemical composition of the litter, the total inflow of N to the forest floor can be estimated. The input of N that will later be stored in stable humus can be calculated by combining the limit value with the linear relationship between N concentration and accumulated mass loss (cf. Sect. 5.4). The N concentration at the limit value can be calculated from the initial concentration of N in litter, and the linear relationship between litter mass loss and litter N concentration (cf. Fig. 11.5). The resulting value is the concentration of N in the remaining stable humus-near litter. This means that we have a tool to calculate the annual accumulation of N in a stable form. This tool also enables us to predict a future N concentration in humus after changes in the ecosystem, for example a change in the tree species

This method has been validated using data from a set of plots with humus that had accumulated for between 2984 and 120 years (Berg and Dise 200Xb). For a first-generation Scots pine forest in a nutrient-poor system (Table 11.1), this way of calculating humus N gave a value of 18 g m^{-2} of N stored after 120 years, compared to an observed value of 15 g m^{-2}. For three systems with humus accumulating for 1106, 2081 and 2984 years, 163, 460, and 761 g m^{-2}, respectively, have been recorded (Table 11.1). A calculation using the above approach (Berg and Dise 200Xb) resulted in calculated amounts of 213, 453, and 677 g m^{-2}, respectively, for the three systems.

The fixation of ammonia may be important in the course of litter decomposition. Some of the N may be bound through chemical reactions including, in part, decomposed lignin and ammonia (Nömmik and Vahtras 1982). This reaction is pH dependent, and it has been shown that N concentration is limiting to the reaction rate, with more free N giving a faster reaction but with higher total N concentration in the litter resulting in a lower reaction rate (Axelsson and Berg 1988). This suggests that in more N-rich litter, N is sequestered through the fixation process at a lower rate. The rate may be related to the number of available reactive sites in the litter material, but this is an unsupported hypothesis at present (Berg et al. 1999b; Berg and Dise 200Xb).

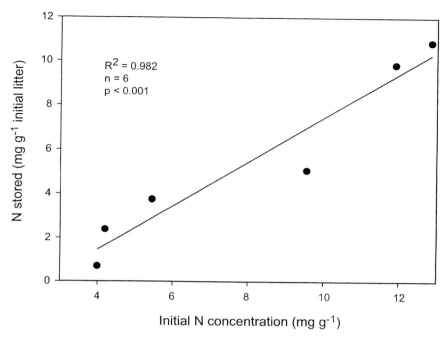

Fig. 11.7. A relationship between litter initial N concentrations and the estimated fraction of N left in the litter, namely the potential capacity to store N in humues. (See potential N storage, Table 11.5) (Berg and Dise 200Xa)

Part of the N in newly shed litter is tied to lignin, and more N is bound during the course of decomposition. Flaig et al. (1959) investigated this process for different kinds of lignin. Berg and Theander (1984) and Berg (1988) found that for Scots pine needle litter, the concentration of N in the "lignin fraction" increased linearly with accumulated mass loss over the course of decomposition.

McClaugherty (unpubl.; Fig. 11.8) followed the amount of N in the lignin fraction for four foliar litters over 2 years of decay. All litters showed an absolute increase in the amount of N bound to lignin during the first year of decay, nearly doubling on the average. Sugar maple and red oak leaves retained their additional lignin-associated N during the second year, but the two conifers, white pine and Canadian hemlock, lost some of their share, with white pine falling back to its initial amount. It is noteworthy that the accumulation of N in the lignin fraction occurred during the first summer of incubation.

Dinitrogen (N_2) fixation has been noted in decaying logs, and may be of importance in the N dynamics of wood decay. Using acetylene reduction to estimate the potential for N_2 fixation, Larsen et al. (1978) found that brown-rotted wood was more likely than white-rotted wood to support N_2 fixation. They also showed differences between species, with Douglas-fir having higher rates of N_2 fixation than subalpine fir or western hemlock. Jurgensen et al. (1984) and Griffiths et al. (1993) found that N_2-fixation potential increased as decay proceeded.

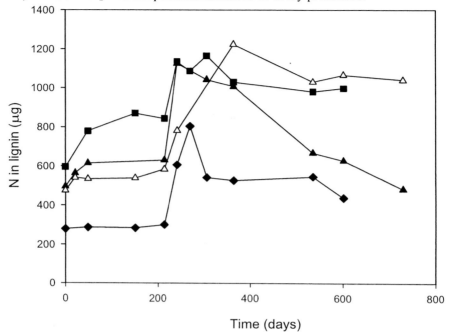

Fig. 11.8. Amount of N associated with the lignin fraction (micrograms per gram initial organic matter content) for four foliar litters incubated for 2 years in a sugar maple forest in Wisconsin, USA. (◆) Sugar maple leaves, (■) red oak leaves, (▲) white pine needles, (△) Canadian hemlock needles (C. McClaugherty unpubl.)

11.5 A climate - change scenario

11.5.1 Fennoscandia and the Baltic basin

A change in climate, leading to increased temperatures and precipitation for boreal regions, would have clear effects on the decomposition patterns of litter, and the accumulation of humus. We have selected a case study for Fennoscandia and the Baltic region that may also be applicable to other boreal regions.

Berg and Meentemeyer (2002) chose to apply a climate-change scenario with an increase in average annual temperature of 4 °C, and an increase in precipitation of 40%, both evenly distributed over the year. Based on that scenario, they calculated AET at a set of different sites, and obtained an average increase in AET of 27%, with minor variation about the mean. Because these forested boreal systems are energy limited, this resulted in the change in AET being rather constant over the whole region.

11.5.2 Is there a general relationship between climate and foliar litter N concentration under boreal/temperate conditions?

Foliar litter fall is richer in N under warmer and wetter conditions

Several factors could be influenced by a climate change, for example: litter chemical composition within species; plant community composition leading to litter with a different chemical composition (cf. Chap. 4); the amount of litter fall (Berg and Meentemeyer 2001).

Within a given species, needle litter will have a higher N concentration with increasing AET (Fig. 4.4). This finding is based on empirical data only (Berg et al. 1995a). Further, available data indicate that a change in species following an increase in temperature and precipitation would lead to more N-rich litter (Berg and Meentemeyer 2001). For example, a change from pine to either spruce or birch, or a change from spruce to birch, would lead to more N-rich litter (Tables 4.4 and 4.5). Even when these species grow on the same soil, and under the same climate, the N concentration would be in the order birch > spruce > pine. In addition, when we compare available data for the N concentration in foliar litter over boreal and temperate forests in Europe to AET, a general relationship indicates that higher AET was related to an increased N concentration (Fig. 4.6). A synthesis of available data for Europe (Berg and Meentemeyer 2001; Fig. 4.6) showed the change in foliar litter N concentration with AET for Scots pine, Norway spruce and deciduous litter.

If we accept the above relationships, which indicate that foliar litters formed at sites with higher AET will have a generally higher N content (Figs. 4.4 and 4.6), this means that such litter would reach a lower limit value during decomposition (Fig. 6.10), and would leave a larger amount of stable material as residue. This would only occur if concentrations of other nutrients that influence the limit value do not change enough to offset the influence of N.

Combining data on estimated limit values, litter N concentrations and AET estimated for a number of boreal and some temperate forest sites, Berg and Meentemeyer (2002) analysed the correlation between limit values and AET. For data on Scots pine only (Fig. 11.9A), this relationship was highly significant. This indicated that within the range of AET values resulting from a the calculated change in climate (from the range of 380-520 to that of 460-650 mm), the limit values fell from nearly 90% to less than 80% decomposition, implying a doubling of the humus accumulation rate due to the decrease in limit values.

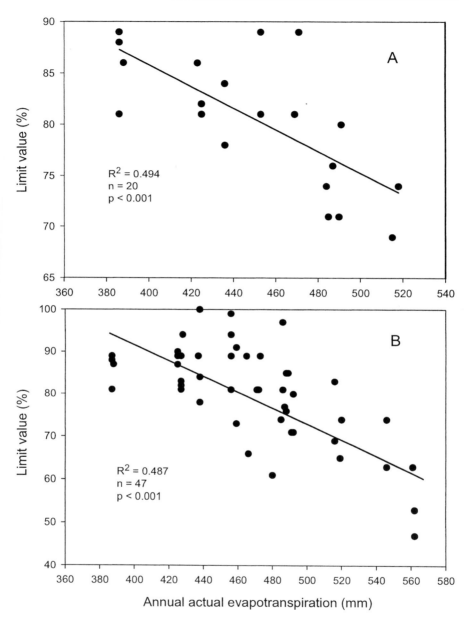

Fig. 11.9. A relationship between limit value for litter decomposition and the climate index AET. The litter originated in all cases from the site at which decomposition was studied. A basis for this relationship is that higher AET would give a foliar litter with higher N concentration, in its turn giving lower limit value (Fig. 11.3). **A** Scots pine litter decomposing at sites along Sweden. **B** Available data for foliar litter decomposing at sites over Europe. (Berg and Meentemeyer 2002)

Litter N concentrations increase faster relative to litter mass loss under warmer and wetter conditions.

Over a large group of litter species, and for litter collected over a broad region, the initial N concentration was not important in regulating the rate of increase in N concentration as litter approaches humus (cf. Chap. 5). However, the significant positive relationships between AET and rate of increase of N concentrations were found for several species including Scots pine, Norway spruce, deciduous species and for analyses across species. When a combination of all available data (coniferous and deciduous litter types) was investigated, the relationship was highly significant (R^2_{adj}=0.628, n=59, p<0.001). Deciduous litters departed from the pattern exhibited by coniferous litters, but the overall relationship remained sigifiusanlimate, as measured by AET, seemed to be a significant factor in affecting the rate of N concentration increase in decomposing leaf litter. These increases were calculated based on accumulated mass loss, which means that, after a particular mass loss, a litter decaying in an area with higher AET would contain more N than if it had decayed in an area with lower AET. The effect of a warmer climate on both the initial N level and the increase rate in N concentration leads to a lower limit value.

11.5.3 How large can a change in humus accumulation be?

A case study at boreal sites. Berg and Meentemeyer (2002) describe a Scots pine site where the AET value is predicted to increase from 470 to 588 mm after the full climate change assumed above. Such a change in climate would still allow the present pine species to grow well at the site, but other species, such as Norway spruce or silver birch, could also be grown there and are likely be planted to have a more productive species at that site.

In order to examine the effect of a change in the substrate quality on humus accumulation, an arbitrary value for foliar litter fall of 2000 kg ha^{-1} was used. Such an assumption is not entirely correct, since a higher litter fall could be a further consequence of a change in the climate (Berg and Meentemeyer 2001). For a Scots pine stand, the increased AET resulted in an increase in the annual humus growth, from 418 to 640 kg ha^{-1}, which can be ascribed to a change in substrate quality, in particular, a higher N concentration (Table 11.7). Using the function (Fig. 11.9B) for all available data, which would also imply a change in species, the annual increase in humus accumulation from 588 to 838 kg ha^{-1} year^{-1} was entirely due to the increase in N concentration.

Including the effect of an increased litter fall for Scots pine at our example stand (Fig. 11.10) would mean an 80% increase in needle litter fall, or 3600 kg ha^{-1} instead of the 2000. With an increase in N concentration leading to a changed limit value (68% instead of 79%) that would mean that the humus accumulation rate increased from 418 kg ha^{-1} year^{-1} to 1152 kg ha^{-1} year^{-1} or ca. 280%. This is a

Table 11.7. An estimate of potential annual increase in humus layers using functions based on Scots pine data only and all available European data. This comparison used an example of a Scots pine site with an AET of 470 mm that after a climate change increased to 588 mm and an annual litter fall of 2000 kg ha^{-1} that increased to 3600 kg ha^{-1}. (In part from-Berg and Meentemeyer 2002)

AET	Limit value	SOM accumulation	Relative increase
[mm]	[%]	[kg ha^{-1} year^{-1}]	[%]
Effect of substrate quality only[a]			
Scots pine data			
470	79.1	418	
588	68.0	640	54
All available European data			
470	79.4	418	
588	58.1	838	100
Effect of substrate quality[a] and increased litter fall			
Scots pine data			
470	79.1	418	
588	68.0	1152	280

[a]Increased N concentration.

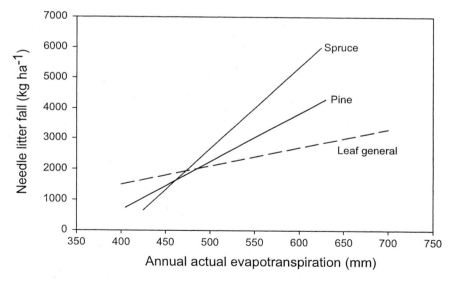

Fig. 11.10. A comparison of needle litter fall to actual evapotranspiration (AET) in monocultural forests of Scots pine and Norway spruce. Part of a relationship found by Meentemeyer et al. (1982) for deciduous leaf litter is indicated by a *dashed line*. (Berg and Meentemeyer 2001)

theoretical exercise, and we should be aware that there may be system influences we are unaware of that might alter the results.

We are not sure what the boundary might be for limit levels, but there must be one. Neither do we have much data from mull soils or from subtropical and tropical systems from which to develop quantitative relationships. Although the quantitative relationships we have observed for boreal and temperate forests may not extend to more tropical climates, the concept of a limit value must still exist, for in the tropics, soil C can be dated to thousands of years in age. Although surface litter may disappear rapidly, C can be stabilized in colloidal forms deeper in the soil.

The estimated reservoir of C in the terrestrial soils of the earth is immense, with at least 1500×10^{15} g present, according to Schlesinger and Andrews (2000), and even a small change in that reservoir could have measurable effects on the atmospheric C concentration.

12 Human activities that influence decomposition

12.1 Introduction

Human activities impact the entire planet. Global climate change and thinning of the stratospheric ozone layer are two anthropogenic factors with worldwide impact that may also influence decomposition of plant litter. On a more continental scale, acid precipitation, including both wet and dry deposition of S and N, is diminishing, but remains important regionally. Ozone is often elevated in urban and industrial areas with possible impacts on decomposition and other ecosystem processes. Also heavy metals that are released into the environment due to human activity, may be detrimental to microorganisms. Some human activities are intended to increase productivity of agricultural and forest ecosystems, and the associated practices can have an impact on decomposition. There is an immense variety of activities that are undertaken in the pursuit of agriculture and forestry, ranging from pesticide application to prescribed burning, activities that to different extents may influence the decomposition of plant litter.

As has become clear throughout this book, decomposition is regulated primarily by factors that influence the ability of the microbial community to process organic matter. These effects may be exerted directly on the microbes, or indirectly, through modification of the substrates undergoing decomposition. Human activities that alter microbial processing will have an effect on the decomposition process. As environmental protection measures reduce many pollutants, decomposition processes are likely to be influenced in places where pollution has been diminished.

Most human activities influence decomposition rates through a modification of climatic or other environmental factors. The climatic effects on decomposition have been covered extensively in Chapter 7, so our discussion of global warming in this chapter will be limited. Similarly, we reviewed the role of N as a regulator of decomposition in Chapter 6, so here we only discuss the role of atmospheric N deposition on decomposition. Just as there are other natural factors that influence decomposition (Chap. 8), there is a variety of anthropogenic factors that may be locally important, and we have chosen only a few of them to offer as examples.

12.2 Global warming

Global warming is an active area of research today, and is the subject of both scientific and political debate. We have described in some detail the role of climate in regulating decomposition (Chap. 7), and the following review of global warming experiments should be considered in the broader context of climatic effects.

Scenarios for global climate change often predict an average increase of 5 °C in the middle latitudes, with smaller changes near the equator and larger changes nearer the poles. Initially, one would predict that global warming would increase soil temperatures, which would increase microbial respiration, resulting in the release of more CO_2 into the atmosphere, providing positive feedback for global warming (Bolin et al. 1986; Intergovernmental Panel on Climate Change 2001).

To study the effect of a potential increase in temperatures in the field, soil-warming experiments were set up in which buried heating cables warmed soils in field plots (Van Cleve et al. 1990). These studies have the distinct advantage of having an undisturbed control, and can provide a longer-term study than most laboratory studies. Still, they have not been long-term enough to consider feedback mechanisms between climate change and the ecosystem. For example, a change in the litter substrate, such as higher nutrient levels, might result from long-term warming (cf. Chap. 4).

Soil warming experiments have indicated that the soil microflora will adapt to higher temperatures, *viz.* a higher temperature will not result in a higher activity (Liski et al. 1999). Indeed, Giardina and Ryan's (2000) work summarizes data from over 80 plots on five continents to show the remarkable insensitivity to changes in temperature of decomposition rates for forest soils. Other studies, including laboratory experiments (Dalias et al. 2001), indicate that the substrate itself changes in composition when initially degraded at elevated temperatures. Both effects counteract an increase in soil temperature. An additional effect of a warmer and wetter climate would be an increased nutrient concentration (N for example) in the newly shed litter, both within (Berg et al. 1995a) and between species (Berg and Meentemeyer 2002). The raised N levels would probably stimulate the decomposition initially, but from the later stages and onwards would inhibit decomposition. In conclusion, the mechanisms that we can distinguish today all indicate that an increase in temperature and precipitation would not result in an increased decomposition rate, but rather an unchanged or lower rate, and would possibly result in an increased level of C storage (Berg et al. 1993a).

Predicting the effects of global warming on decomposition requires an understanding of both moisture and temperature dynamics. Timing and amount of precipitation can be as important as changes in temperature. If warming does increase initial rates of decay, but has little effect on the decay of more stable organic matter (e.g. litter in the late stage of decay), the important question becomes whether warming influences the amount of stable organic matter ultimately produced. For example, if accelerated initial decomposition increases N availability, then N-stabilized humus may actually increase. Clearly, this is an area in need of further investigation.

12.3 Regional pollution

Industrial society inevitably creates by-products that can be considered as pollutants, in the sense that the substances are present in concentrations, and at locations, where they can damage biological processes. Some of these pollutants may be found in significant concentrations relatively near to their source, such as heavy metals from a smelter. Others may be derived from widespread processes, such as fossil fuel combustion, and be found over larger regions. For example, the substances that give rise to acid precipitation come from a variety of anthropogenic sources, and may spread over areas of continental scale.

12.3.1 Atmospheric N and S deposition

Atmospheric N and S deposition has been the focus of ecological research in North America and western Europe for several decades. The possible effects are numerous. To some extent, N and S inputs could actually serve as fertilizers. In some areas, the amount of N deposition has been large enough that N is no longer a primary limiting factor for ecosystem productivity, a condition known as N saturation (Aber et al. 1998). Under N saturation, decomposition could be influenced by several different mechanisms (Jefferies and Maron 1997). A common form of input is as oxides of N and S that are readily converted to acids. The acid can influence the pH of the soil system, and therefore alter mobility, and thus availability, of nutrients. This in turn could influence both the composition and the functioning of the microbial community.

Sulfur dioxide. A significant set of studies on the effects of SO_2 deposition on decomposition examined the effect of this pollutant on leaf litter decomposition rates (Newsham et al. 1992a), on the fungal community (Newsham et al. 1992b), and on the chemical content of the litter (Newsham et al. 1992c). In their study on fungal communities, Newsham et al. (1992b) used both an existing gradient of SO_2 and an experimental open-air field-fumigation system. They incubated six different species of leaf litter, and isolated and identified the invading fungi. In general, SO_2 did not influence the rate of fungal invasion into the litter, but it did change the species composition of the community. As SO_2 concentrations increased, sensitive fungal species disappeared and were replaced by more tolerant species. The next question is whether the change in composition was accompanied by a change in function.

The impact of SO_2 pollution on leaf litter mass loss rates was generally small and not significant, though there was a trend towards lower mass loss with time under heavy SO_2 fumigation (Newsham et al. 1992b). In a parallel study, they inoculated four species of fungi onto sterile leaves and observed respiration under control and elevated SO_2. Only one of the four species showed a decline in respiration due to fumigation with SO_2.

An indirect effect of SO_2 on decomposition would occur if the litter chemistry were altered. In their third study, Newsham et al. (1992c) compared the chemical content of leaf litters formed in the presence of SO_2 pollution. Increased atmospheric SO_2 resulted in greater losses of Ca and Mg from the litter and a lower litter pH. Although both Ca and Mg are important for fungal growth and metabolism, it is not clear whether the leaching was sufficient to retard the activity of the fungal community. In this group of studies, decomposition measured as respiration, was suppressed under high SO_2, but mass loss was not. In a different study, using mixed deciduous litter and Scots pine needle litter (Wookey and Ineson 1991; Wookey et al. 1991), very similar results were achieved. Increased SO_2 decreased respiration, increased efflux of Ca and Mg from litter, and caused either a very small reduction, or even a positive effect on litter mass loss.

Although SO_2 is an antimicrobial agent at very high concentrations, it does not seem to have a large or consistent effect on decomposition at levels of pollution normally encountered. Fritze et al. (1992) studied decomposition-related phenomena along a transect downwind from a coastal oil refinery. They found no effect of pollution on soil respiration, litter mass loss from Scots pine needles or the C and N content of microbial biomass. The only notable effect was a decrease in length of fungal hyphae in more polluted areas of the transect.

Nitrogen deposition. Nitrogen is being added to ecosystems around the globe at unprecedented levels, but the long-term ecological implications of this N enrichment remain uncertain (Jefferies and Maron 1997). There are three principal ways that N deposition might influence decomposition. First, N enrichment could alter litter chemistry by changing concentrations and ratios of nutrients, and perhaps also by influencing the organic chemical composition of the litter. These changes could occur within existing species, or could be influenced by a change in the community composition in response to enrichment. Second, N deposition could influence the microbial community. As described earlier, N levels in the environment may have an effect on the lignin degrading fungi (cf. Chap. 3), but the direction of the effect may depend on the level of enrichment, and the nature of the existing fungal community. Third, N enrichment may influence the kinetics of humus formation and the nature of the compounds formed.

Several studies have utilized low-level chronic additions of N to study the effect of N deposition on ecosystem processes including decomposition. The results are variable. The NITREX project, a series of large-scale N saturation experiments conducted at seven sites in western Europe (Wright et al. 1995), yielded several studies on the response of decomposition to increased N deposition. Emmett et al. (1995) found that adding N at two to four times the background level of 17 kg N ha^{-1} $year^{-1}$ over a period of 17 months had no effect on the mass loss of Sitka spruce needles incubated in litter bags on the forest floor, up to a mass loss of ca. 55%. The litter used in this study was not produced under N enrichment and was thus uniform among treatments. Boxman et al. (1998) studied the decomposition of forest floor (L and F) material at three of the NITREX sites. They found large differences in decay rates among the three sites, but no differences related to the level of experimental N enrichment. The litters used were not grown under N enrichment. Simultaneous enrichment with N and S did cause a decline in mass loss rates at an alpine NITREX site in Norway (Wright and Tietema 1995). In this study, the direct effect of the treatment on decomposition of heather leaf litter was separated from the indirect effect mediated through modification in litter quality, using reciprocal transplants of litter. Combined N and S enrichments slowed decay through both direct and indirect mechanisms. The experiment did not separate the effects of N and S.

To test the importance of N enrichment on decomposition, Magill and Aber (1998) incubated four species of foliar litter in two different forest types, a red pine plantation and a mixed northern hardwood stand dominated by black and red oaks. Both stands were located at Harvard Forest (see Appendix III). There were three levels of N enrichment in each forest: control (0 g N m^{-2} $year^{-1}$), low N (5 g N m^{-2} $year^{-1}$), and high N (15 g N m^{-2} $year^{-1}$), applied in six equal applications per year during the growing season, and continuing for 6 years. The lowest mass loss occurred in the high N stands; the control stands had the greatest mass loss, though this was not much greater than the low N treated stands. A general conclusion from this experiment is that increased exogenous N availability can reduce long-term decay rates, but may have little effect on early stages of decay.

Acid precipitation. Acid precipitation received considerable research attention during the 1970s and 1980s (Hutchinson and Havas 1980). Although the amount of acid precipitation has declined, it does continue, and there may be a legacy from earlier, higher, amounts of deposition. Kuperman (1999) studied leaf litter decomposition along a gradient of long-term bulk deposition across the Ohio Valley, increasing from Illinois through Indiana to Ohio (USA), which was described by Armentano and Loucks (1990). After 19 months, including two growing seasons, mean mass loss, expressed as a percent of initial dry matter, of white oak leaves was 36% in Illinois, 54% in Indiana and 70% in Ohio. Dynamics of N, P and S followed mass loss patterns, with the greatest amount of all three nutrients released in Ohio, and the least in Illinois. The sites were described as having similar climates. It appears that N and S deposition may favor decomposition during the first 2 years of decay. How this might influence humus formation in these systems remains unclear, but higher N availability can increase the amount of humus stabilized (see Chap. 11).

12.3.2 Heavy metals

Heavy metals, including Ni, Cu, Zn, Pb, Fe and Cd, are often dispersed in the environment in association with industrial activities. Low concentrations of these elements occur naturally, and can be found in decomposing substrates (Laskowski and Berg 1993). These elements vary in their toxicity, mobility and potential to alter decomposition patterns. Nevertheless, research has often found increased accumulation of litter in areas contaminated with heavy metals (Strojan 1978, Coughtrey et al. 1979, Freedman and Hutchinson 1980). Since these early studies, others have examined in detail the role of heavy metals in decomposition. It is possible that the effect of heavy metals will vary with stage of decomposition and litter type. Thus, for Scots pine needle litter, Berg et al. (1991) found only a weak suppression in the early stage, whereas in the late stage that require degradation of lignin, a clear trend towards a considerably slower degradation was seen. Berg et al. (1991) suggested that the lignin degradation was more sensitive to disturbances, although the exact reasons remain to be discovered.

During litter decomposition, most heavy metals are retained in the litter structure, and their concentrations increase, often exponentially. Berg and Laskowski (1993) described the decomposition of natural, unpolluted needle litter, and found that the concentration of heavy metals increased to levels that had been clearly inhibitory in a pollution transect.

12.3.3 Nuclear radiation, ozone, and proximity to urban centers

Radiation. Gamma irradiation is not commonly encountered in nature, but studies in France and the USA have examined its effects on numerous ecosystem properties. Woodwell and Marples (1968) examined the effects of chronic gamma irradiation on humus and litter decay rates in an oak-pine forest in New York State, USA. Irradiation had two effects on the system; a decline in leaf litter fall, and a slight increase in decay rates of litter. From an ecosystem perspective, the major effect was more on altering the amount and types of litter than on modification of the decomposition process. As irradiation continued, leaf litter fall declined, but input of woody litter from dying vegetation increased.

Poinsot-Balaguer and Tabone (1995) studied decomposition in an irradiation study that had been ongoing for 16 years. In their study, decomposition was much lower closer to the radiation source. Lower decay rates were related to lower population densities of microarthropods. Both of these results could be attributed almost entirely to water relations. Close to the source, larger woody vegetation had been killed and the remaining vegetation, comprised mostly of annuals, was sparse and of low stature, leading to strong reductions in soil moisture. Thus, in both these studies on gamma irradiation, it appears that radiation strong enough to kill vegetation had much less of an effect on decomposition than did moisture and litter chemical quality.

Ozone. Ozone appears to have little direct or indirect effect on decomposition, in spite of the fact that it has a wide variety of effects on both the physiology and biochemistry of organisms. The coniferous forests of the San Bernadino Mountains, near Los Angeles, California (USA), were visibly damaged by air pollution during the 1970s. Studies indicated an ozone gradient decreasing from west to east. Fenn and Dunn (1989) studied litter from this gradient and found that CO_2 efflux from needles in the L-layer incubated under laboratory conditions was greatest from litter collected at the most polluted sites. In addition, fungal diversity and N content of the litter were greater at the more polluted sites. The possibility that ozone physically damaged the needles in a manner that favored their decomposition was rejected in another study (Fenn 1991), in which damaged and undamaged Ponderosa pine needles were incubated simultaneously and exhibited no differences. The author rules out climatic influences and suggested that these results may be due to increased N deposition that is parallel to ozone concentrations. Litter organic chemical composition was not determined in this study.

In an eastern deciduous forest, Boerner and Rebbeck (1995) grew seedlings of three species of forest trees under three levels of ozone (none, ambient, double-ambient), and then studied mass loss and N dynamics in a 1-year litter bag study. Mass loss was not influenced by the ozone concentration in which the foliage grew. Ozone increased the initial nitrogen release rate in sugar maple and yellow poplar, but not in black cherry, even though initial N concentrations in the litters were not affected by exposure to ozone.

Urban-rural gradient. Because so many pollutants vary together along an urban-rural gradient, it is difficult to identify individual factors. However, the combined effects may be of significance to decomposition processes. Carreiro et al. (1999) collected red oak leaf litter along an urban to rural gradient beginning in New York City. They subjected that litter to a microbial bioassay to determine its decomposability. After 150 days of incubation in laboratory microcosms, there were significant differences in mass loss, with 34.3% for rural, 31.1% for suburban and 25.9% for urban. The lignin-to-N ratio and the concentration of hemicellulose were significantly lower at rural sites, but there were no other measured differences in litter composition. Indicators of fungal growth were greatest at rural sites, but bacterial indicators were similar across all sites. Although this study only covers the early stage of decay, it does demonstrate an influence of proximity to urban environments on initial litter quality, and thus on initial decay rates. Longer-term effects have yet to be described.

12.4 Effects of selected forest management practices

Many studies have been carried out on the effects of thinning, clear-cutting, soil compaction, site preparation and other management practices on decomposition and nutrient cycling in forests. The variables involved are often hard to control or measure and the results differ from site to site, making it difficult to extract generalizations. Our purpose here is not to review the literature, but to merely provide a few sample studies.

Gadgil and Gadgil (1978) performed a particularly interesting study that demonstrates the complexity of studies on management practices and decomposition. They found that the presence of plants actually slowed the decomposition of Monterey pine needle litter by means of mycorrhizal suppression of decomposition. Clearcutting, by eliminating mycorrhizae in the soil, released the litter from the suppression and allowed a more rapid rate of mass loss. Their experimental design controlled variations in both soil moisture and temperature. Their view is in contrast to the more general view that clear-felling modifies decomposition primarily by altering the site's climate. This finding is not general however. Cortina and Valejo (1994) studied Monterey pine needle litter decomposition in Spain and found that clear-felling resulted in a large decrease in mass loss relative to a control stand, with first year mass losses of 25 and 37%, respectively.

Different litter types respond differently to clear-cutting. In British Columbia, Canada, Prescott et al. (2000) studied the decomposition of lodgepole pine and trembling aspen foliar litters in forests and adjacent clearcuts. After 3 years of decomposition, the mass loss for aspen was the same between forested and clear-cut areas. For lodgepole pine, mass loss was significantly ($p<0.05$) greater in the forest than the clear-cut after 1 year, and the differences remained significant throughout the 4 years of the study. The mass loss after 4 years, averaged over 16 stands, was 66% in the forest and 60% in the clear-cut.

Whole tree harvesting removes large amounts of organic matter from the system and can thus influence the environment for decomposition. Kranabetter and Chapman (1999) investigated the effects of three levels of harvesting (stem-only, whole-tree and whole tree plus removal or scalping of the forest floor), and two levels of soil compaction, on decomposition of leaf litter in trembling aspen stands in British Columbia. They found that the highest mass loss rates occurred with whole tree harvesting and soil compaction; the lowest rates occurred with the two most extreme practices: stem only harvesting with no compaction and whole tree plus forest floor removal with compaction. They suggested that microclimatic differences might be able to explain the different decomposition rates.

Prescribed burning is sometimes used in forest management to control fuel buildup and to stimulate nutrient cycling. Monleon and Cromack (1996) investigated the effect of prior burning on Ponderosa pine needle litter decomposition. Using a series of sites which had been burned 3 months, 5 years and 12 years before the study, they found that in all cases, decomposition, measured over an 18-month period, was reduced on previously burned sites. They suggest that the differences were due more to changing the near-ground environment than to altering nutrient dynamics.

Although many more examples could be described, there is no simple summary to be derived. There are simply so many site-specific factors that can influence decomposition, that simple generalizations about the effects of given management practices are not possible. Only if the effect of a management practice can be predicted at the microbial and chemical levels, can the effect on decomposition be understood.

12.5 Long-term perspective

Most of the studies presented here have been relatively short-term, although results from some of these could be extrapolated to obtain mechanisms for humus accumulation and storage. However, we must still assume that long-term storage of humus may be regulated in different ways among forest ecosystems. Wild fires have generally been the main agent depleting the C sinks in forest soils, with lesser amounts of C lost due to harvesting of forests and conversion of land from forest to agricultural or other uses. The current strategy of wildfire prevention, along with the present need to quantify C sinks, and determine their long-term stability, clearly illustrate the need for real long-term research efforts. We expect that such efforts will reveal surprises in terms of storage mechanisms.

Most studies on litter decomposition have been short term, often covering just a fraction of the entire litter-to-humus process. As a consequence, many empirical conclusions and model-based predictions have, of necessity, attempted to extrapolate from information about the early stages, to encompass the whole process, with the result that much of the scientific literature in this field is based on rather incomplete assumptions. There is an urgent requirement for more reliable information about C and N sinks, given the current need for an improved understanding of the global balance of atmospheric CO_2 levels. More long-term approaches and new techniques are required to generate information to increase our knowledge of the interaction between climate change and decomposition processes. Long-term studies that cover the full decomposition process for different litter species and types, which can reveal new facts about sequestration mechanisms, are very much needed. Further, we must regard ecosystems that are composed of different litter species as units with potentially very different properties relating to long-term decomposition and C sequestration. Long-term humus development may also be affected by site history, and techniques that can analyse long-term C sequestration would be invaluable tools.

Appendix I. Glossary

Short definitions of phrases and terms used in the book.

Accumulated mass loss The total amount of mass lost from a decomposing substrate, usually expressed as a percent of initial mass.

Aerobic Oxygen demanding. The term as used here refers to microbial processes which demand the presence of oxygen.

Allophane A soil dominated by amorphous (noncrystalline) clay-sized aluminosilicates. These are frequently found in highly weathered volcanic deposits.

Ammonium/ammonia fixation Fixed NH_3 is the NH_3 that is retained by the soil organic matter or decomposing plant litter after intensive extraction and leaching with either diluted mineral acid or neutral salt solutions [1]. Not to be confused with dinitrogen fixation.

Anaerobic Does not demand oxygen. The term as used here refers to microbial processes that can proceed without the presence of oxygen.

Biomass (1) Organic matter present as live microbial tissue. (2) The mass of organic material produced by living organisms, including both living and nonliving tissues.

Breakdown   This term is used here to indicate "...a reduction in particle size of the organic resource" [2] or **comminution.** A similar effect is brought about by abiotic factors such as freezing and thawing, or wetting and drying cycles.

Constant fractional rate Refers to decay rate according to first-order kinetics (e.g. radioactive decay) in which the decomposition of material proceeds at a constant rate for all periods of the process, until the original material has been completely used up.

Continentality For our purpose the effects of climate continentality concern temperature and precipitation. Thus when comparing temperature and precipitation along a transect from the sea towards inland the annual average precipitation decreases and the mean annual temperature decreases. The temperature amplitudes increase both over the day and over the year, e.g. between July and January.

Decomposition We have used the word "decomposition" or "mass-loss" as the loss of mass from plant litter due to microbial decomposition or leaching of water-soluble substances. Decomposition can also be defined as litter CO_2 release plus leaching of compounds. Breakdown (above) is not included in the concept.

Fulvic acid Colored material which remains in solution after removal [3, 4] of humic acid by acidification.

Humic acid The dark-colored organic material which can be extracted from [3, 4] soil by various reagents and which is insoluble in dilute acid.

Humic substances A series of relatively high-molecular-weight, brown to black-colored [3, 4] substances formed by secondary synthesis reactions. The term is used as a generic name to describe the colored material, or its fractions, obtained on the basis of solubility characteristics. These materials are a distinctive characteristic of the soil (or sediment) environment, in that they are dissimilar to the biopolymers of microorganisms and higher plants (including lignin).

Humin The alkali-insoluble fraction of soil organic matter or humus [3, 4].

Humus Sum total of the stable organic substances in the soil, not including undecayed animal and plant tissues, partially decomposed material and the soil biomass [3, 4].

Leaching The loss of nutrients and incompletely decomposed organic compounds [5] from the intact remains of decomposing litter, due to the action of water.

LH factor Litter-to-humus factor. The remaining fraction of the litter when decomposition has reached the limit value, namely (100-limit value)/100 [6].

Limit value Calculated value for the extent of decomposition of a given litter type, at which the decomposition rate approaches zero.

Litter The same as litter remains (cf. newly shed litter).

Litter CO_2 release The mineralization of carbon from litter. Gives mass loss from litter minus leaching of compounds. See "decomposition".

Litter remains Undecayed plant and animal tissues, and their partial decomposition products [3, 4].

Newly shed litter Plant litter that has been shed so recently that the decomposition processes have not yet started. This is complicated by the presence of newly shed litter which starts decomposing when still attached to the plant.

Non-humic compounds Compounds belonging to known classes of compounds, such as amino acids, carbohydrates, fats, waxes, resins, organic acids, etc. Humus probably contains most, if not all, of the biochemical compounds synthesized by living organisms [3, 4].

Potential humus Is a theoretical concept and refers to the fraction of litter that can be calculated to become humus. It has also been called the LH factor.

Sequestration We have used the term for long-term storage of mainly C and N, and occasionally other nutrients. The use follows the definition given by Webster's Dictionary [7] and refers to the fact that the compounds are bound and removed from the biological activities in their system, unless the system is subject to such changes that it may be considered altered. Our use differs from that currently used by plant physiologists, namely that a compound (e.g. C) bound into live plant tissue is sequestered. The binding of e.g. CO_2 to plant tissue is one of several steps in the process of sequestration and only part of the C taken up by plants is sequestered, the rest being released as CO_2 during decomposition.

Soil organic matter The same definition as for humus [3, 4].

Steady state A term sometimes used also about humus layers. We have not found any real definition of the term directed to humus but only a suggestion to an application for a specific boreal region of country size. Thus Schulze et al. [8] suggested that today's humus layers, accumulated after the last glaciation, reflect a steady state, namely the amount that has accumulated considering all possible influencing factors, including fire and anthropogenic influences. A problem with such a definition is of course that when e.g. forest fires are suppressed, like today, the basic conditions for the definition are changed. On a smaller geographical scale, e.g. a stand level, it would not be useful. In this book we have avoided to use the term 'steady state' for humus. The reason is that we have not found any evidence for the validity of such a concept. We cannot exclude, though, that in ecosystems with developing humus layers, steady states do exist.

References

[1] Nömmik and Vahtras (1982)
[2] Swift et al. (1979)
[3] Stevenson (1982)
[4] Waksman (1936)
[5] This vol.
[6] Berg et al. (2001)
[7] Gove (1996)
[8] Schultze et al. (1989)

Appendix II. Scientific names of vascular plants

We have listed here the vascular plant species mentioned in the text. Where the same species has different common names in American and European English, we have given both, indicated with (A) and (E), respectively, followed by the Latin/botanical name. Different dictionaries give different common names for the same species, and our purpose here has been to give the correct common names as they were used here, rather than to list the most widely accepted. Plants are divided into two groups: gymnosperms and angiosperms. Within each group, species and genera are arranged alphabetically by the American common name.

II.1 Gymnosperms

Firs

Douglas-fir (A, E) *(Pseudotsuga menziesii* Mirb. Franco.) (=*Pseudotsuga douglasii)*
European silver fir (A), common silver fir (E) *(Abies alba)* (=*Abies pechinata* D.C.)
Pacific silver fir (A, E) *(Abies amabilis* Douglas ex J. Forbes)
Subalpine fir (A) *(Abies lasiocarpa* (Hook.) Nutt.)

Hemlocks

Eastern hemlock (A), Canadian hemlock (E), *(Tsuga canadensis* (L.) Carr.)
Western hemlock (A,E), *(Tsuga heterophylla* (Raf.) Sarg.)

Pines

Austrian pine (A, E), *(Pinus nigra* Arnold)
Corsican pine (A, E), *(Pinus nigra* var. *maritima)*
Eastern white pine, white pine (A), Weymoth pine (E) *(Pinus strobus* L.)
Jack pine (A, E) *(Pinus banksiana* Lamb.)
Limber pine (A, E) *(Pinus flexilis* James)
Loblolly pine (A, E) *(Pinus taeda* L.)
Lodgepole pine (A, E) *(Pinus contorta* var. *latifolia* Engelm.)
Lodgepole pine (A), shore pine (E) *(Pinus contorta* var *contorta)*
Maritime pine (E) *(Pinus pinaster* Ait.)
Norway pine (A), red pine (A, E) *(Pinus resinosa* Ait.)

Ponderosa pine (A), western yellow pine (E) (*Pinus ponderosa* Laws.)
Scots or Scotch pine (A, E) (*Pinus sylvestris* L)
Stone pine (E) (*Pinus pinea* L.)

Spruces

Norway spruce (A, E) (*Picea abies* (L.) Karst.) (=*Picea excelsa* Link.)
Sitka spruce (A, E) (*Picea sitchensis* (Bong.) Carr)
Red spruce (A, E) (*Picea rubens* Sarg.) (=P. rubra [DuRoi] Link)
White spruce (A, E) (*Picea glauca* (Moench.) Voss)

II.2 Angiosperms

Alders

Red alder (A), Oregon alder (E) (*Alnus rubra* Bong.) (=*A.oregona* Nutt.)
Gray alder (A), grey alder (E) (*Alnus incana* (L.) Moench)
Nepalese alder (A, E) (*Alnus nepalensis* D.Don)

Aspen

Bigtooth aspen (A, E) (*Populus grandidenta* Michx.)
Quaking aspen, trembling aspen (A, E) (*Populus tremuloides* Michx.)

Beeches

American beech (A) (*Fagus grandifolia* Ehrh.)
European beech (A), common beech (E) (*Fagus silvatica* L.)
Japanese beech (A, E) (*Fagus crenata* Bl.)

Birches

Black birch, sweet birch (A) (*Betula lenta* L.)
European white birch (A); common birch, silver birch, weeping birch, white birch (E)
        (*Betula pendula* Roth. = B. *verrucosa* Ehrh.)
Hairy birch (A), downy birch (E) (*Betula pubescens* Ehrh.)
Yellow birch (A) (*Betula alleghaniensis* Britt.) (= B. *lutea*)

Grasses

Perennial ryegrass, English ryegrass (A), Italian ryegrass (E) (*Lolium multiflorum*
        Lam.) (=*L. perenne* var. *multiflorum* (Lam.) Parnell)

Small six-weeks grass (A) (*Vulpia microstachys* (Nutt.) Munro).
Soft chess (A), soft brome (A,E) *(Bromus hordaceus* L. = *B. mollis* L.*)*
Wild oats (A,E) *(Avena fatua* L.*)*

Maples

European maple (A), Norway maple (A, E) (*Acer platanoides* L.)
Red maple (A, E) (*Acer rubrum* L.)
Sugar maple (A, E) (*Acer saccharum* Marsh.)

Oaks

Black oak (A) (*Quercus velutina* Lam.)
Chestnut oak (A), basket oak (E) (*Quercus prinus* L.)
Common oak, pedunculate oak (E), English oak (A) (*Quercus robur* L.)
Durmast oak (E) (*Quercus petraea* (Mattuschka) Lieblein)
Eastern red oak, northern red oak (A), Red oak (E) (*Quercus rubra* L.) (*Q. rubra* du
    Roi) (*Q borealis* Michx. f.)
Pyrenean oak (E) (*Quercus pyrenaeica* Willd.) (=*Q. toza* D.C.)
White oak (A), American white oak (E) (*Quercus alba* L.)

Other woody plants

Black cherry (A, E) (*Prunus serotina* Ehrh..)
European ash (A), common ash (E) (*Fraxinus excelsior* L.)
European blueberry(A), Bilberry (E) (*Vaccinium myrtillus* L)
European mountain ash, mountain ash (A), Rowan (E) (*Sorbus aucuparia* L.)
Filbert (A), common hazel (E) (*Corylus avellana* L.)
Flowering dogwood (A), cornel (E) (*Cornus florida* L.)
Heather (A, E) (*Calluna vulgaris* (L.) Hull)
Lingonberry (A), cowberry (E), (*Vaccinium vitis-idea* L.)
Sierra palm (A), mountain palm (E) (*Prestoea montana* (R. Graham) Nichols.)
Tabonuco, gommier, candle tree (A, E) (*Dacryodes excelsa* Vahl)
Yellow poplar, tulip poplar (A, E) (*Liriodendron tulipfera* L.)

Appendix III. Site descriptions

Stråsan, Sweden

An optimum-nutrition experiment was established at Stråsan, central Sweden (60°55′N; 16°01′E), on a till soil. The site is located on a steep, west-facing slope, at a mean altitude of 350 m. The annual mean temperature is 3.4 °C, the annual precipitation 775 mm and the average AET is 422 mm. The area was planted with Norway spruce in 1958. For the fertilization experiment, started in 1967, the area was divided into 30×30-m plots. A detailed description of the experiment is given by Tamm et al. (1974). Fertilizer was added once annually, with the dosages for the medium dosag plots (N2P2) given in Table III.1. Solid fertilizer was dispensed by hand (ammonium nitrate and superphosphate). During the course of the decomposition experiment, fertilizer was applied in early May in 1982, 1983, 1984, 1985 and 1986. In early May, 1984, K, Mg, Mn, B, Zn, Cu and Mo were added as well.

Jädraås, Sweden

Jädraås (Site no 6:51) in central Sweden, about 200 km north of Stockholm, has a Scots pine monoculture. The site is sometimes called Ih 5. It has a forest about 130 years old (in 1980) located at 60°49′N;16°30′E and at an altitude of 185 m. The forest is situated on a very nutrient-poor sediment soil. The annual mean precipitation is 609 mm and the long-term average temperature is 3.8 °C. The ground vegetation is composed mainly of European blueberry, cowberry, heather, mosses and lichens. The humus form is mor, and the soil profile a podzol. The soil texture is fine sand. Additional information on this site is provided by Axelsson and Bråkenhielm (1980).

Black Hawk Island, Wisconsin, USA

Black Hawk Island is a 70-ha island located in the Wisconsin River, near Wisconsin Dells, at 43°40′N and 89°45′W, and 260-280 m in elevation. Mean annual precipitation is 800 mm and AET is 605 mm. The island is mostly covered with old growth forest, and includes stands dominated by each of the following species:

sugar maple, red oak, white oak, Canadian hemlock, white pine, red pine, and big-tooth aspen. Soils were highly variable across the site, and included Entisols, Spodosols, Alfisols, Inceptisols and Histosols. The vegetation and soils are de-scribed in greater detail by Pastor et al. (1982)

Harvard Forest, Massachusetts, USA

Harvard Forest is located in Petersham, Massachusetts, USA (42°30'N and 72°10'W), and is part of the Long Term Ecological Research (LTER) network. The two sites referred to in this book are both located in the Prospect Hill tract at altitudes between 340 and 360 m. The red pine stand was planted in 1925, and the second growth mixed hardwood stand was last cut in about 1900. The mixed hardwood stand was dominated by red oak and red maple, with lesser amounts of black birch and American beech. Forest floors in both sites were mors. Soils were very stony Inceptisols, of glacial origin. Average annual precipitation was 1120 mm and AET 580 mm. Mean annual temperature was 7 °C; mean monthly temperatures range from 19 °C in July to -12 °C in January. N deposition (wet + dry) is about 8 kg ha^{-1} year^{-1}.

Table III.1. Dosages of fertilizer applied to the fertiolized (N2P2) plots in the optimum nutrition experiment at Stråsan. No fertilizer was applied to control (N0) plots. Table data from Tamm et al. (1974) and from A. Aronsson (pers. comm.)

Year	Dosage applied [kg ha^{-1} year^{-1}]	
	N	P
1967	120	40
1968	120	
1969	120	40
1970	80	20
1971	80	
1972	80	
1973	80	
1974	80	40
1975	80	
1976	80	
1977	60	40
1978	60	
1979	60	
1980	60	40
1981	60	
1982	60	
1983	60	
1984	60	40
1985	60	
1986	60	

References

Aber JD, Melillo JM (1982) Nitrogen immobilization in decaying hardwood leaf litter as a function of initial nitrogen and lignin content. Can J Bot 60:2263-2269

Aber JD, McClaugherty CA, Melillo JM (1984) Litter decomposition in Wisconsin forests - Mass loss, organic-chemical constituents and nitrogen. Univ Wisc Res Bull R3284, University of Wisconsin, Madison, WI

Aber JD, Melillo JM, McClaugherty CA (1990) Predicting long-term patterns of mass loss, nitrogen dynamics, and soil organic matter formation from initial fine litter chemistry in temperate forest ecosystems. Can J Bot 68:2201-2208

Aber JD, McDowell WH, Nadelhoffer KJ, Magill A, Bernston G, Kamakea M, McNulty SG, Currie W, Rustad L, Fernandez I (1998) Nitrogen saturation in temperate forest ecosystems: hypotheses revisited. BioScience 48:921-934

Aerts R (1997) Climate, leaf litter chemistry and leaf-litter decomposition in terrestrial ecosystems - a triangular relationship. Oikos 79:439-449

Agee JK, Huff MH (1987) Fuel succession in western hemlock - Douglas-fir forest. Can J For Res 17:697-704

Ågren G, Bosatta E, (1996) Quality: a bridge between theory and experiment in soil organic matter studies. Oikos 76:522-528

Alban DH, Pastor J (1993) Decomposition of aspen, spruce, and pine boles on two sites in Minnesota. Can J For Res 23:1744-1749

Anagnost SE (1998) Light microscopic diagnosis of wood decay. IAWA J 19:141-167

Ander P, Eriksson K-E (1977) Selective degradation of wood components by white-rot fungi. Physiol Plant 41:239-248

Ander P, Marzullo L (1997) Sugar oxidoreductases and veratryl alcohol oxidase as related to lignin degradation. J Biotech 53:115-131

Anderson JM, Macfadyen A (eds) (1976) The role of terrestrial and aquatic organisms in decomposition processes. Blackwell, Oxford, 474 pp

Anonymous (1996) Forstliche Standortsaufnahme. Begriffe, Definitionen, Einteilungen, Kennzeichnungen, Erläuterungen, 5th edn. IHW-Verlag, Eching. ISBN 3-930167-18-2 (in German)

Anonymous (2002) Atlas Florae Europaeae issued by the Botanical Museum of the Finnish Museum of Natural History, Helsinki. Web site http://www.fmnh.helsinki.fi/map/-afe/E_afe.htm?pageid=571&language=English

Archibald F, Roy B (1992) Production of manganic chelates by laccase from the lignin-degrading fungus *Trametes (Coriolus) versicolor*. Appl Environ Microbiol 58:1496-1499

Armentano T, Loucks O (1990) Spatial patterns of S and N deposition in the midwestern hardwoods region. In: Loucks O (ed) Air pollutants and forest response: The Ohio Corridor Study. Year 4 Annual Report. Miami University, Holcomb Research Institute, Oxford, Ohio

Arthur MA, Fahey TJ (1990) Mass and nutrient content of decaying boles in an Engelmann spruce/subalpine fir forest, Rocky Mountain National Park, Colorado. Can J For Res 20:730-727

Arthur MA, Tritton LM, Fahey TJ (1993) Dead bole mass and nutrients remaining 23 years after clear-felling of a northern hardwood forest. Can J For Res 23:1298-1305

Axelsson G, Berg B (1988) Fixation of ammonia (^{15}N) to Scots pine needle litter in different stages of decomposition. Scand J For Res 3:273-280

Axelsson B, Bråkenhielm (1980) Investigation sites of the Swedish Coniferous Forest Project - biological and physiological features. In: Structure and function of northern coniferous forest. An ecosystem study. Ecol Bull (Stockh) 32:391-400

Bakken LR (1997) Cultural and nonculturable bacteria in soil. In: Van Elsas JD, Trevors JT, Wellington MH (eds) Modern soil microbiology. Dekker, New York, pp 47-62

Bal L (1973) Micromorphological analysis of soils. Soil Survey Papers of the Netherlands Soil Survey Institute No 6, Wageningen

Beatty SW, Stone EL (1986) The variety of soil micro-sites created by tree falls. Can J For Res 16:539-548

Bélaich JP, Tardif C, Bélaich A, Caudin C (1997) The cellulolytic system of *Clostridium cellulolyticum*. J Biotech 57:3-14

Bengtsson G (1992) Interactions between fungl, bacteria, and beech leaves in a stream microcosm. Oecologia 89:542-549

Berg B (1984) Decomposition of root litter and some factors regulating the process: long-term root litter decomposition in a Scots pine forest. Soil Biol Biochem 16:609-617

Berg B (1988) Dynamics of nitrogen (^{15}N) in decomposing Scots pine (*Pinus silvestris* L.) needle litter. Long-term decomposition in a Scots pine forest VI. Can J Bot 66:1539-1546

Berg B (1998a) A maximum limit for foliar litter decomposition - a synthesis of data from forest systems. Dept For Ecol For Soil, Swed Univ Agric Sci. Rept 77, 158 pp

Berg B (1998b) Organic-matter quality and C/N ratio as controlling factors of RSOM turnover. In: Refractory soil organic matter (RSOM): structure and stability. Bayreuth, 26-28 Apr 1998. Mitt Dtsch Bodenkundl Ges 87:79-91

Berg B (2000a) Initial rates and limit values for decomposition of Scots pine and Norway spruce needle litter - a synthesis for N-fertilized forest stands. Can J For Res 30:122-135

Berg B (2000b) Litter decomposition and organic matter turnover in northern forest soils. For Ecol Manage 133:13-22

Berg B, Cortina J (1995) Nutrient dynamics in some decomposing leaf and needle litters in a Pinus sylvestris forest. Scand J For Res 10:1-11

Berg B, Dise N (200Xa) Does a higher N content in plant litter give a higher N sequestration in soil organic matter? Water, Air Soil Pollut (submitted)

Berg B, Dise N (200Xb) Calculating the long-term stable nitrogen sink in northern Eurpean forests Glob Change Biol (submitted)

Berg B, Ekbohm G (1983) Nitrogen immobilization to decomposing needle litter at variable carbon - nitrogen ratios. Ecology 64:63-67

Berg B, Ekbohm G (1991) Litter mass loss rates and decomposition patterns in some needle and leaf litter types. Long-term decomposition in a Scots pine forest VII. Can J Bot 69:1449-1456

Berg B, Ekbohm G (1993) Decomposing needle litter in lodgepole pine (*Pinus contorta*) and Scots pine (*Pinus sylvestris*) monocultural systems. Is there a maximum mass loss? Scand J For Res 8:457-465

Berg B, Johansson M-B (1998) A maximum limit for foliar litter decomposition - a synthesis of data from forest systems. Dept Forest Ecology Forest Soils, Swed Univ Agric Sci Rep 77, 158 pp

Berg B, Lundmark J-E (1987) Decomposition of needle litter in lodgepole pine and Scots pine monocultures - a comparison. Scand J For Res 2:3-12

Berg B, Matzner E (1997) The effect of N deposition on the mineralization of C from plant litter and humus. Environ Rev 5:1-25

Berg B, McClaugherty C (1989) Nitrogen and phosphorus release from decomposing litter in relation to the disappearance of lignin. Can J Bot 67:1148-1156

Berg B, Meentemeyer V (2001) Litterfall in European pine and spruce forests as related to climate. Can J For Res 31:292-301

Berg B, Meentemeyer V (2002) Litter quality in a north European transect versus carbon storage potential. Plant Soil 242:83-92

Berg B, Staaf H (1980a) Decomposition rate and chemical changes in Scots pine needle litter. II. Influence of chemical composition. Ecol Bull (Stockh) 32:373-390

Berg B, Staaf H (1980b) Leaching, accumulation and release of nitrogen from decomposing forest litter. Ecol Bull (Stockh) 32:163-178

Berg B, Staaf H (1981) Chemical composition of main plant litter components at Ivantjärnsheden - data from decomposition studies. Swed Conif For Proj Int Rept 104, 10 pp

Berg B, Staaf H (1983) The influence of slash removal on soil organic matter and nutrients in a Scots pine forest soil I & II; II. Short-term dynamics of carbon and nitrogen pools in soil and the contents of plant nutrients in the forest floor. Swed Conif For Proj Tech Rep 34:25-66

Berg B, Staaf H (1987) Release of nutrients from decomposing white birch leaves and Scots pine needle litter. Pedobiologia 30:55-63

Berg B, Tamm CO (1991) Decomposition and nutrient dynamics of litter in long-term optimum nutrition experiments. I. Organic matter decomposition in Norway spruce (*Picea abies*) needle litter. Scand J For Res 6:305-321

Berg B, Theander O (1984) The dynamics of some nitrogen fractions in decomposing Scots pine needles. Pedobiologia 27:161-167

Berg B, von Hofsten B, Pettersson G (1972) Electron microscopic observations on the degradation of cellulose fibres by *Cellvibrio fulvus* and *Sporocytophaga myxococcoides*. J Appl Bact 35:215-219

Berg B, Hannus K, Popoff T, Theander O (1980) Chemical components of Scots pine needles and needle litter and inhibition of fungal species by extractives. In: Persson T (ed) Structure and function of northern coniferous forests - an ecosystem study. Ecol Bull (Stockh) 32:391-400

Berg B, Hannus K, Popoff T, Theander O (1982a) Changes in organic-chemical components during decomposition. Long-term decomposition in a Scots pine forest I. Can J Bot 60:1310-1319

Berg B, Wessen B, Ekbohm G (1982b) Nitrogen level and lignin decomposition in Scots pine needle litter. Oikos 38:291-296

Berg B, Jansson PE, Meentemeyer V (1984) Litter decomposition and climate - regional and local models. In: Ågren G (ed) State and change of forest ecosystems - indicators in current research. Swed Univ Agric Sci, Dept Ecol Environ Res Rep 13:389-404

Berg B, Ekbohm G, McClaugherty CA (1984) Lignin and holocellulose relations during long-term decomposition of some forest litters. Long-term decomposition in a Scots pine forest. IV. Can J Bot 62:2540-2550

Berg B, Staaf H, Wessén B (1987) Decomposition and nutrient release in needle litter from nitrogen-fertilized Scots pine (*Pinus sylvestris*) stands. Scand J For Res 2:399-415

Berg B, Booltink HGW, Breymeyer A, Ewertsson A, Gallardo A, Holm B, Johansson MB, Koivuoja S, Meentemeyer V, Nyman P, Olofsson J, Pettersson AS, Staaf H, Staaf I, Uba L (1991a) Data on needle litter decomposition and soil climate as well as site characteristics for some coniferous forest sites, 2nd edn, sect 2, Data on needle litter decomposition. Swed Univ Agric Sci, Dept Ecol Environ Res Rep 42, 450 pp

Berg B, Ekbohm G, Söderström B, Staaf H (1991b) Reduction of decomposition rates of Scots pine needle litter due to heavy metal pollution. Water Air Soil Pollut 69:165-177

Berg B, Berg M, Bottner P, Box E, Breymeyer A, Calvo de Anta R, Couteaux M, Gallardo A, Escudero A, Kratz W, Madeira M, Mälkönen E, MeentemeyerV, Muñoz F, Piussi P, Remacle J, Virzo De Santo A (1993a) Litter mass loss in pine forests of Europe and eastern United States as compared to actual evapotranspiration on a European scale. Biogeochemistry 20:127-153

Berg B, Berg M, Bottner P, Box E, Breymeyer A, Calvo de Anta R, Couteaux M, Gallardo A, Escudero A, Kratz W, Madeira M, Meentemeyer V, Muñoz F, Piussi P, Remacle J, Virzo De Santo A (1993b) Litter mass loss in pine forests of Europe: relationships with climate and litter quality. In: Breymeyer A (ed) Proc SCOPE Seminar. Conf Pap 18. Geography of Carbon Budget Processes in Terrestrial Ecosystems. Szymbark, 17-23 Aug 1991, pp 81-110

Berg B, Berg M, Cortina J, Escudero A, Gallardo A, Johansson M, Madeira M (1993c) Soil organic matter in some European coniferous forests. In: Breymeyer A (ed) Proc SCOPE Seminar. Conf Pap 18. Geography of carbon budget processes in terrestrial ecosystems. Szymbark, 17-23 Aug 1991, pp 111-122

Berg B, Berg M, Flower-Ellis JGK, Gallardo A, Johansson M, Lundmark J-E,Madeira M (1993d) Amounts of litterfall in some European Coniferous Forests. In: Breymeyer A (ed) Proc SCOPE Seminar. Conf Pap 18. Geography of carbon budget processes in terrestrial ecosystems. Szymbark, 17-23 Aug 1991, pp 123-146

Berg B, McClaugherty C, Johansson M (1993e) Litter mass-loss rates in late stages of decomposition at some climatically and nutritionally different pine sites. Long-term decomposition in a Scots pine forest VIII. Can J Bot 71:680-692

Berg B, Calvo de Anta R, Escudero A, Johansson M-B, Laskowski R, Madeira M, McClaugherty C, Meentemeyer V, Reurslag A, Virzo De Santo A (1995a) The chemical composition of newly shed needle litter of different pine species and Scots pine in a climatic transect. Long-term decomposition in a Scots pine forest X. Can J Bot 73:1423-1435

Berg B, McClaugherty C, Virzo De Santo A, Johansson M-B, Ekbohm G (1995b) Decomposition of forest litter and soil organic matter - a mechanism for soil organic matter buildup? Scand J For Res 10:108-119

Berg B, Johansson M-B, Lundmark J-E (1996a) Uppbyggnad av organiskt material i skogsmark - har gödsling och trädslagsval en inverkan? In: Berg B (ed) Markdagen 1996. Dept For Ecol For Soil, Swed Univ Agric Sci Rep 72:33-44 (in Swedish)

Berg B, Ekbohm G, Johansson M-B, McClaugherty C, Rutigliano F, Virzo De Santo A (1996b) Some foliar litter types have a maximum limit for decomposition - a synthesis of data from forest systems. Can J Bot 74:659-672

Berg B, McClaugherty C, Johansson M-B (1997) Chemical changes in decomposing plant litter can be systemized with respect to the litter's initial chemical composition. Dept For Ecol For Soil, Swed Univ Agric Sci Rep 74, 85 pp

Berg B, Meentemeyer V, Johansson M-B, Kratz W (1998) Decomposition of tree root litter in a climatic transect of forests in northern Europe - a synthesis. Scand J For Res 13:402-412

Berg B, Albrektson A, Berg M, Cortina J, Johansson M-B, Gallardo A, Madeira M, Pausas J, Kratz W, Vallejo R, McClaugherty C (1999a) Amounts of litterfall in pine forests in the northern hemisphere, especially Scots pine. Ann For Sci 56:625-639

Berg B, Laskowski R, Virzo De Santo A (1999b) Estimated N concentration in humus as based on initial N concentration in foliar litter - a synthesis. Can J Bot 77:1712-1722

Berg B, Meentemeyer V, Johansson M-B (2000) Litter decomposition in a climatic transect of Norway spruce forests - climate and lignin control of mass-loss rates. Can J For Res 30:1136-1147

Berg B, McClaugherty C, Virzo De Santo A, Johnson D (2001) Humus buildup in boreal forests - effects of litter fall and its N concentration. Can J For Res 31:988-998

Berg B, Virzo De Santo A, Rutigliano F, Ekbohm G (200X) Limit values for plant litter decomposing in two contrasting soils – influence of litter elemental composition. Acta Oecologia (submitted)

Bergkvist B (1986) Metal fluxes in Spruce and Beech forest ecosystems of south Sweden. PhD Thesis, Univ Lund, Sweden

Bethge PO, Rådeström R, Theander O (1971) Kvantitativ kolhydratbestämning - en detaljstudie. Comm Swed For Prod Res Lab 63B. SE-114 86 Stockholm, 48 pp (in Swedish)

Blair JM (1988a) Nitrogen, sulphur and phosphorus dynamics in decomposing deciduous leaf litter in the southern Appalachians. Soil Biol Biochem 20:693-701

Blair JM (1988b) Nutrient release from decomposing foliar litter of three tree species with special reference to calcium, magnesium and potassium dynamics. Plant Soil 110:49-55

Blair JM, Parmelee RW, Beare MH (1990) Decay rates, nitrogen fluxes, and decomposer communities of single- and mixed-species foliar litter. Ecology 71:1976-1985

Blanchette RA (1984) Manganese accumulation in wood decayed by white rot fungi. Phytopathology 74:725-730

Blanchette RA (1991) Delignification by wood-decay fungi. Annu Rev Phytopathol 29:381-398

Blanchette RA (1995) Degradation of the lignocellulose complex in wood. Can J Bot 73 [Suppl]:S999-S1010

Blanchette RA, Shaw CG (1978) Associations among bacteria, yeasts, and basidiomycetes during wood decay. Phytopathology 68:631-637

Blanchette RA, Krueger EW, Haight JE, Akhtar M, Akin DE (1997) Cell wall alterations in loblolly pine wood decayed by the white-rot fungus Ceriporiopsis subvermispora. J Biotech 53:203-213

Bloomfield J, Vogt KA, Vogt DJ (1993) Decay rate and substrate quality of fine roots and foliage of two tropical tree species in the Luquillo Experimental Forest, Puerto Rico. Plant Soil 150:233-245

Bockheim JG, Leide LE (1986) Litter and forest floor dynamics in Wisconsin. Plant Soil 96:393-406

Bocock KL, Gilbert O, Capstick CK, Twinn DC, Waid JS, Woodman MJ (1960) Changes in leaf litter when placed on the surface of soils with contrasting humus types. I. Losses in dry weight of oak and ash leaf litter. J Soil Sci 11:1-9

Boddy L, Watkinson SC (1995) Wood decomposition, higher fungi, and their role in nutrient redistribution. Can J Bot 73 [Suppl 1]:S1377-S1383

Boddy L, Owens EM, Chapela IH (1989) Small scale variation in decay rate within logs one year after felling: effect of fungal community structure and moisture content. FEMS Microbiol Ecol 62:173-184

Boerner REJ, Rebbeck J (1995) Decomposition and nitrogen release from leaves of three hardwood species grown under elevated O_3 and/or CO_2. In: Collins HP, Robertson GP, Klug MJ (eds) The significance and regulation of soil biodiversity. Kluwer, Amsterdam, pp 169-177

Bogatyrev L, Berg B, Staaf H (1983) Leaching of plant nutrients and total phenolic substances from some foliage litters - a laboratory study. Swed Conif For Proj Tech Rep 33, 59 pp

Bolin B, Doos BR, Jager J, Warrick R (eds) (1986) The greenhouse effect, climatic change and ecosystems. SCOPE 29. Wiley, Chichester, 541 pp

Bollag JM, Minard RD, Liu SY (1983) Cross-linkage between anilines and phenolic constituents. Environ Sci Technol 17:72-80

Bono JJ, Gas G, Boudet M, Fayret J, Delatour C (1984) Etude comparee de la degradation de lignocelluloses par differentes souches de *Fomes annosus*. Can J Microbiol 29:1683-1688

Bormann BT, DeBell DS (1981) Nitrogen content and other soil properties to age of red alder stands. Soil Sci Soc Am J 45:428-432

Bosatta E, Ågren G (1985) Theoretical analysis of decomposition of heterogeneous substrates. Soil Biol Biochem 17:601-610

Bosatta E, Ågren G (1991) Dynamics of carbon and nitrogen in the organic matter of the soil: a general theory. Am Nat 138:227-245

Boxman AW, Blanck K, Brandrud TE, Emmett BA, Gundersen P, Hogervorst RF, Kjønaas OJ, Persson H, Timmermann V (1998) Vegetation and soil biota response to experimentally-changed nitrogen inputs in coniferous forest ecosystems of the NITREX project. For Ecol Manage 101:65-79

Bray MW, Andrews TM (1924) Chemical changes of groundwood during decay. Indus Eng Chem 16(2):137-139

Bremer E, van Houtum W, van Kessel C (1991) Carbon dioxide evolution from wheat and lentil residues affected by grinding, added nitrogen, and absence of soil. Biol Fertil Soils 11:221-227

Breymeyer A, Laskowski R (1999) Ecosystem process studies along a climatic transect at 52-53° N, 12-32° E: pine litter decomposition. Geog Polon 72:45-64

Bringmark E, Bringmark L (1991) Large-scale pattern of mor layer degradation in Sweden measured as standardized respiration. In: Humic substances in the aquatic and terrestrial environments. Proceedings of an international symposium, Linköping, Sweden, 21-23 Aug 1989. Lecture notes in earth sciences, no 33. Springer, Berlin Heidelberg New York, pp 255-259

Broadbent FE, Stevenson FJ (1966) Organic matter Interactions. In: McVickar HN et al (eds) Agricultural anhydrous ammonia. Technology and use. Proc Symp St Louis, Mo, 29-30 Sept 1965. Agric Ammonia Inst, Memphis, Tenn; Am Soc Agron and Soil Sci Soc Am, Madison, WI, pp 169-187

Bunell F, Tait DEN, Flanagan PW, Van Cleve K (1977) Microbial respiration and substrate weight loss. I. A general model of the influence of abiotic variables. Soil Biol Biochem 9:33-40

Burke MK, Raynal DJ (1994) Fine root growth phenology, production, and turnover in a northern hardwood forest ecosystem. Plant Soil 162:135-146

Cameron MD, Aust SD (2001) Cellobiose dehydrogenase - an extracellular fungal flavocytochrome. Enzyme Microbiol Tech 28:129-138

Camiré C, Côté B, Brulotte S (1991) Decomposition of roots of black alder and hybrid poplar in short rotation plantings: nitrogen and lignin control. Plant Soil 138:123-132

Carpenter SR (1981) Decay of heterogeneous detritus: a general model. J Theor Biol 89:539-547

Carreiro MM, Howe K, Parkhurst DF, Pouyat RV (1999) Variation in quality and decomposability of red oak leaf litter along an urban rural gradient. Biol Fertil Soils 30:258-268

Chadwick DR, Ineson P, Woods C, Piearce TG (1998) Decomposition of *Pinus sylvestris* litter in litter bags: Influence of underlying native litter. Soil Biol Biochem 30:47-55

Chambers JQ, Higuchi N, Schimel JP, Ferreira LV, Melak JM (2000) Decomposition and carbon cycling of dead trees in tropical forests of the central Amazon. Oecologia 122:380-388

Cole DW, Compton JE, Edmonds RL, Homann PS, Van Miegroet H (1995) Comparison of carbon accumulation in douglas fir and red alder forests. In: McFee WW, Kelly JM (eds) Carbon forms and functions in Forest Soils. Soil Sci Soc Am, Madison, WI, USA, pp 527-546

Core HA, Cote WA, Day AC (1979) Wood structure and identification. Syracuse Univ Press, Syracuse, 182 pp

Cotrufo MF, Ineson P (1995) Effects of enhanced atmospheric CO_2 and nutrient supply on the quality and subsequent decomposition of fine roots of *Betula pendula* Roth. and *Picea sitchensis* (Bong.) Carr. Plant Soil 170:P267-P277

Cotrufo MF, Ineson P (1996) Elevated CO_2 reduces field decomposition rates of *Betula pendula* (Roth.) leaf litter. Oecologia 106:525-530

Cotrufo F, Berg B, Kratz W (1998) Increased atmospheric CO_2 concentrations and litter quality. Environ Rev 6:1-12

Couteaux M-M, McTiernan K, Berg B, Szuberla D, Dardennes P (1998) Chemical composition and carbon mineralisation potential of Scots pine needles at different stages of decomposition. Soil Biol Biochem 30:583-595

Coughtrey PJ, Jones CH, Martin MH, Shales SW (1979) Litter accumulation in woodlands contaminated by Pb, Zn Cd and Cu. Oecologia 39:51-60

Crawford RL (1981) Lignin biodegradation and transformation. Wiley, New York, 137 pp

Dahlgren RA, Vogt KA, Ugolini FC (1991) The influence of soil chemistry on fine root aluminum concentrations and root dynamics in a sub alpine Spodosol, Washington State, USA. Plant Soil 133:117-129

Dalias P, Anderson JM, Bottner P, Coûteaux MM (2001) Long-term effects of temperature on carbon mineralisation processes. Soil Biol Biochem 33:1049-1057

David MB, Vance GF, Rissing JM, Stevenson FJ (1989) Organic carbon fractions in extracts of O and B horizons from a New England Spodosol: effects of acid treatment. J Environ Qual 18:212-217

Dean JFD (1997) Lignin analysis, chap 17. In: Dashek WV (ed) Methods in plant biochemistry and molecular biology. CRC Press, New York, pp 199-215

De Haan S (1977) Humus, its formation, its relation with the mineral part of the soil and its significance for soil productivity. In: Organic matter studies, vol 1. International Atomic Agency, Vienna, pp 21-30

Dekker RFH (1985) Biodegradation of hemicelluloses. In: Higuchi T (ed) Biosynthesis and biodegradation of wood components. Academic Press, Tokyo, pp 505-533

Delaney M, Brown S, Lugo AE, Torres-Lezama A, Quintero NB (1998) The quantity and turnover of dead wood in permanent forest plots in six life zones of Venezuela. Biotropica 30(1):2-11

Delcourt HR, Harris WF (1980) Carbon Budget of the South-Eastern US Biota: analysis of historical change in trend from source to sink. Science 210:321-323

D'Souza TM, Merritt CS, Reddy CA (1999) Lignin-modifying enzymes of the white-rot basidiomycete *Ganoderma lucidum*. Appl Environ Microbiol 65:5307-5313

Dugger WM (1983) Boron in plant metabolism. In: Läuckli A, Bielski RL (eds) Inorganic plant nutrition, vol 15B. Springer, Berlin Heidelberg New York, pp 626-650

Dwyer LM, Merriam G (1981) Influence of topographic heterogeneity on deciduous litter decomposition. Oikos 37:228-237

Dwyer LM, Merriam G (1983) Decomposition of natural litter mixtures in a deciduous forest. Can J Bot 62:2340-2344

Dyer ML (1986) A model of organic decomposition rates based on climate and litter properties. MA Thesis, Univ Georgia, Athens, GA, 78 pp

Dyer ML, Meentemeyer V, Berg B (1990) Apparent controls of mass loss rate of leaf litter on a regional scale. Scand J For Res 5:311-323

Dziadowiec H (1987) The decomposition of plant litter fall in an oak-linden-hornbeam forest and oak-pine mixed forest of the Bialowieza National Park. Acta Soc Bot Polon 56:169-185

Edmonds RL (1979) Decomposition and nutrient release in Douglas-fir needle litter in relation to stand development. Can J For Res 9:132-140

Effland MJ (1977) Modified procedure to determine acid insoluble lignin in wood and pulp. Tech Assoc Pulp Pap Ind J 60(10):143-144

Emmett BA, Brittain SA, Hughes S, Kennedy V (1995) Nitrogen additions (NaNO$_3$ and NH$_4$NO$_3$) at Aber Forest, Wales: II. Response of trees and soil nitrogen transformations. For Ecol Manage 71:61-73

Erickson HE, Edmonds RL, Peterson CE (1985) Decomposition of logging residues in Douglas-fir, western hemlock, Pacific silver fir, and ponderosa pine ecosystems. Can J For Res 15:914-921

Eriksson K-E, Blanchette RA, Ander P (1990) Microbial and enzymatic degradation of wood and wood components. Springer, Berlin Heidelberg New York, 407 pp

Fahey TJ, Hughes JW (1994) Fine root dynamics in a northern hardwood forest ecosystem, Hubbard Brook Experimental Forest, NH. J Ecol 82:533-548

Fahey TJ, Hughes JW, Pu M, Arthur M (1988) Root decomposition and nutrient flux following whole-tree harvest of northern hardwood forest. For Sci 34:744-768

Faituri MY (2002) Soil organic matter in Mediterranean and Scandinavian forest ecosystems and dynamics of nutrients and monomeric phenolic compounds. Silvestra 236, 136 pp

Fengel D, Wegener G (1983) Wood: chemistry, ultrastructure, reactions. De Gruyter, Berlin, 613 pp

Fenn M (1991) Increased site fertility and litter decomposition rate in high-pollution sites in the San Bernadino mountains. For Sci 37:1163-1181

Fenn ME, Dunn PH (1989) Litter decomposition across an air-pollution gradient in the San Bernardino mountains. Soil Sci Soc Am J 53:1560-1567

Flaig W, Schobinger U, Deuel H (1959) Umwandlung von Lignin in Huminsäuren bei der Verrottung von Weizenstrah. Chem Ber 92:1973-1982 (in German)

Flessa H, Ludwig B, Heil B, Merbach W (2000) The origin of soil organic C, dissolved organic C and respiration in a long-term maize experiment in Halle, Germany, determined by ^{13}C natural abundance. J Plant Nutr Soil Sci 163:157-163

Flower-Ellis J (1985) Litterfall in an age series of Scots pine stands: summary of results for the period 1973-1983. Dept Ecol Environ Res. Swed Univ Agric Sci Rept 19:75-94

Fog K (1988) The effect of added nitrogen on the rate of decomposition of organic matter. Biol Rev 63:433-462

Fogel R, Cromack K (1977) Effect of habitat and substrate quality on Douglas fir litter decomposition in western Oregon. Can J Bot 55:1632-1640

Forrest WG, Ovington JD (1970) Organic matter changes in an age series of *Pinus radiata* plantations. J Appl Ecol 7:110-120

Franck VM, Hungate BA, Chapin FS, Field CB (1997) Decomposition of litter produced under elevated CO_2: dependence on plant species and nutrient supply. Biogeochemistry 36:223-237

Freedman B, Hutchinson TC (1980) Effects of smelter pollution on forest leaf litter decomposition near a nickel-copper smelter at Sudbury, Ontario. Can J Bot 58:1722-1736

Freer SN, Detroy RW (1982) Biological delignification of ^{14}C-labeled lignocelluloses by basidiomycetes: degradation and solubilization of the lignin and cellulose components. Mycologia 74:943-951

Fritze H, Kiikkilä O, Pasanen J, Pietikäinen J (1992) Reaction of forest soil microflora to environmental stress along a moderate pollution gradient next to an oil refinery. Plant Soil 140:175-182

Gadgil RL, Gadgil PD (1978) Influence of clearfelling on decomposition of *Pinus radiata* litter. N Z J Sci 8:213-224

Gholz HL, Wedin DA, Smitherman SM, Harmon ME, Parton WJ (2000) Long-term dynamics of pine and hardwood litter in contrasting environments: toward a global model of decomposition. Glob Change Bio 6:751-765

Giardina CP, Ryan MG (2000) Evidence the decomposition rates of organic carbon in mineral soil do not vary with temperature. Nature 404:858-861

Gilbertson RL (1980) Wood-rotting fungi of North America. Mycologia 72:1-49

Gilliam FS, Yurish BM, Adams MB (2001) Temporal and spatial variation of nitrogen transformations in nitrogen-saturated soils of a central Appalachian hardwood forest. Can J For Res 31:1768-1785

Goodburn JM, Lorimer CG (1998) Cavity trees and coarse woody debris in old-growth and managed northern hardwood forests in Wisconsin and Michigan. Can J For Res 28:427-438

Gore JA, Patterson WA (1986) Mass of downed wood in northern hardwood forests in New Hampshire: potential effects of forest management. Can J For Res 16:335-339

Gosz JR (1981) Nitrogen cycling in coniferous ecosystems. In: Clark FE, Rosswall T (eds) Terrestrial nitrogen cycles. Processes, ecosystem strategies, and management impacts. Ecol Bull (Stockh) 33:405-426

Gove PB ed. (1996) Webster's New Encyclopedic Dictionary, New revised edition. Könemann Verlag, Köln. 1637 pp

Graham RL, Cromack K (1982) Mass, nutrient content, and decay rate of dead boles in rain forests of Olympic National Park. Can J For Res 17:304-310

Green F, Highley TL (1997) Mechanisms of brown-rot decay: paradigm or paradox. Int Biodet Biodeg 39:113-124

Griffith GS, Boddy L (1990) Fungal decomposition of attached angiosperm twigs. I. Decay community development in ash, beech, and oak. New Phytol 116:407-415

Griffiths R, Caldwell BA, Cromack K, Morita RY (1990) Douglas-fir forest soils colonized by ectomycorrhizal mats. 1. Seasonal variation in nitrogen chemistry and nitrogen transformation rates. Can J For Res 20:211-218

Griffiths RP, Harmon ME, Caldwell BA, Carpenter SE (1993) Acetylene reduction in conifer logs during early stages of decomposition. Plant Soil 148:53-61

Guggenberger G (1994) Acidification effects of dissolved organic matter mobility in spruce forest ecosystems. Environ Int 20:31-41

Hanson PJ, Edwards NT, Garten CT, Andrews JA (2000) Separating root and soil microbial contributions to soil respiration: a review of methods and observations. Biogeochemistry 48:115-146

Harley JL, Smith SE (1983) Mycorrhizal symbiosis. Academic Press, London, 483 pp

Harmon ME, Sexton J (1996) Guidelines for measurement of woody detritus in forest ecosystems. US Long Term Ecological Research Network, Albuquerque, NM, 42 pp

Harmon ME, Franklin JF, Swanson FJ, Sollins P, Gregory SV, Lattin JD, Anderson NH, Cline SP, Aumen, NG, Sedel JR, Lienkaemper GW, Cromack K, Cummins, KW (1986) Ecology of coarse woody debris in temperate ecosystems. Adv Ecol Res 15:133-302

Harmon ME, Sexton J, Caldwell BA, Carpenter SE (1994) Fungal sporocarp mediated losses of Ca, Fe, K, Mg, Mn, N, P, and Zn from conifer logs in the early stages of decomposition. Can J For Res 24:1883-1893

Harmon ME, Whigham DF, Sexton J, Olmsted I (1995) Decomposition and mass of woody detritus in the dry tropical forests of the northeastern Yucatan peninsula, Mexico. Biotropica 27:305-316

Harmon ME, Krankina ON, Sexton J (2000) Decomposition vectors: a new approach to estimating woody detritus decomposition dynamics. Can J For Res 30:76-84

Hatakka A (2001) Biodegradation of lignin. In: Hofman M and Stein A (eds) Biopolymers, vol 1. Lignin, Humic substances and Coal. Wiley, Weinheim, pp 129-180

Hendrick RL, Pregitzer KS (1992) The demography of fine roots in a northern hardwood forest. Ecology 73:1094-1104

Hendricks JJ, Nadelhoffer KJ, Aber JD (1993) Assessing the role of fine roots in carbon and nutrient cycling. Trend Ecol Evol 8:174-178

Highley TL (1987) Changes in chemical components of hardwood and softwood by brown-rot fungi. Mater Organ 21:39-45

Highley TL (1988) Cellulolytic activity of brown-rot and white-rot fungi on solid media. Holzforschung 42:211-216

Highley TL, Murmanis LL, Palmer JG (1985) Micromorphology of degradation in western hemlock and sweetgum by the brown-rot fungus *Poria placenta*. Holzforschung 39:73-78

Higuchi T (1993) Biodegradation mechanism of lignin by white-rot basidiomycetes. J Biotech 30:1-8

Hintikka V, Näykki O (1967) Notes on the effects of the fungus *Hydnellum ferrugineum* on forest soil and vegetation. Commun Inst For Fenn 62:1-22

Hirano T, Tanaka H, Enoki A (1997) Relationship between production of hydroxyl radicals and degradation of wood by the brown-rot fungus, *Tyromyces palustris*. Holzforschung 51:389-395

Högberg P, Nordgren A, Buchmann N, Taylor A, Ekblad A, Högberg M, Nyberg G, Ottosson-Löfvenius M, Read D (2001) Large-scale forest girdling shows that current photosynthesis drives soil respiration. Nature 411:789-792

Houghton RA (2001) Counting terrestrial sources and sinks of carbon. Clim Change 48:525-534

Howard PJA, Howard DM (1974) Microbial decomposition of tree and shrub leaf litter. Oikos 25:311-352

Howard DM, Howard PJA (1980) Effect of species, source of litter, type of soil, and climate on litter decomposition. Microbial decomposition of tree and shrub leaf litter 3. Oikos 34:115-124

Huang Y, Street-Perrott FA, Metcalfe SE, Brenner M, Moreland M, Freeman KH (2001) Climate change as the dominant control on glacial-interglacial variations in C3 and C4 plant abundances. Science 293:1647-1651

Huhta V, Persson T, Setälä H (1998) Functional implications of soil fauna diversity in boreal forests. Appl Soil Ecol 10:277-288

Hutchinson TC, Havas M (1980) Effects of acid precipitation on terrestrial ecosystems. NATO conference on effects of acid precipitation on vegetation and soils, Toronto, Ontario, 1978. Plenum Press, New York, 654 pp

Intergovernmental Panel on Climate Change (2001) Climate change 2001: the scientific basis. Contribution of working group I to the third assessment report. Cambridge Univ Press, Cambridge, 881 pp

Jansson PE, Berg B (1985) Temporal variation of litter decomposition in relation to simulated soil climate. Long-term decomposition in a Scots pine forest V. Can J Bot 63:1008-1016

Jefferies RL, Maron JL (1997) The embarrassment of riches: atmospheric deposition of nitrogen and community and ecosystem processes. Trends Ecol Evol 12:74-78

Jenkinson DS (1977) Studies on the decomposition of plant material in soil V. The effects of plant cover and soil type on the loss of carbon from ^{14}C labeled ryegrass decomposing under field conditions. J Soil Sci 28:424-434

Jenny H (1980) The soil resource. Origin and behavior. Springer, Berlin Heidelberg New York

Jenny H, Gessel SP, Bingham FT (1949) Comparative study of decomposition rates of organic matter in temperate and tropical regions. Soil Sci 68:419-432

Jin L, Schultz TP, Nicholas DD (1990) Structural characterization of brown-rotted lignin. Holzforschung 44:133-138

Johansson M-B (1987) Radikal markberedning - ett olämpligt sätt att utnyttja kväveförrådet i avverknings-resterna. Sveriges Skogsvårdsförbunds Tidskrift 2:35-41 (in Swedish)

Johansson M-B, Berg B, Meentemeyer V (1995) Litter mass-loss rates in late stages of decomposition in a climatic transect of pine forests. Long-term decomposition in a Scots pine forest. IX. Can J Bot 73:1509-1521

Johnson D, Lindberg SE (1992) Atmospheric deposition and forest nutrient cycling – a synthesis of the integrated forest study. Ecol. Stud. 91

Johnson D, Todd DE (1998) Effects of harvesting intensity on forest productivity and soil carbon storage in a mixed oak forest. In: Lal R, Kimble JR, Follett J, Stewart BA (eds) Management of carbon sequestration in soil. CRC Press, New York, pp 351-364

Jones PCT, Mollison JE (1948) A technique for the quantitative estimation of soil microorganisms. J Gen Microbiol 2:54-69

Jurgensen MF, Larsen MJ, Spano SD, Harvey AE, Gale MR (1984) Nitrogen fixation associated with increased wood decay in Douglas-fir residue. For Sci 30:1038-1044

Kayang H (2001) Fungi and bacterial enzyme activities in *Alnus nepalensis* D. Don. Eur J Soil Sci 37:175-180

Kelsey RG, Harmon ME (1989) Distribution and variation of extractable total phenols and tannins in the logs of four conifers after 1 year on the ground. Can J For Res 19:1030-1036

Kerr TJ, Goring DAI (1975) The ultrastructural arrangement of the wood cell wall. Cell Chem Technol 9:563-573

Keyser P, Kirk TK, Zeikus IG (1978) Ligninolytic enzyme of *Phanerochaete chrysosporium*: synthesized in the absence of lignin in response to nitrogen starvation. J Bacteriol 135:790-797

Kirk TK (1980) Physiology of lignin metabolism by white rot fungi. In: Kirk TK, Higuchi T, Chang H (eds) Lignin biodegradation: microbiology, chemistry, and potential applications, vol 2. CRC Press, Boca Raton, pp 51-63

Kirk TK (1984) Degradation of lignin. In: Gibson DT (ed) Microbial degradation of organic compounds. Marcel Dekker, New York, 399-437.

Knapp EB, Elliott LF, Campbell GS (1983) Microbial respiration and growth during the decay of wheat straw. Soil Biol Biochem 15:319-323

Kögel-Knabner I, Ziegler F, Riederer M, Zech W (1989) Distribution and decomposition pattern of cutin and suberin in forest soils. Z Pflanzenernähr Bodenkd 152:409-413

Kolattukudy PE (1980) Biopolyester membranes of plants: cutin and suberin. Science 208:990-1000

Kolattukudy PE (1981) Structure, biosynthesis, and biodegradation of cutin and suberin. Annu Rev Plant Physiol 32:359-367

Kranabetter JM, Chapman BK (1999) Effects of forest soil compaction and organic matter removal on leaf litter decomposition in central British Columbia. Can J Soil Sci 79:543-550

Krankina ON, Harmon ME, Griazkin AV (1999) Nutrient stores and dynamics of woody debris in a boreal forest: modeling potential implications at the stand level. Can J For Res 29:20-32

Kuperman RG (1999) Litter decomposition and nutrient dynamics in oak-hickory forests along a historic gradient of nitrogen and sulfur deposition. Soil Biol Biochem 31:237-244

Kurz WA, Kimmins JP (1987) Analysis of some sources of error in methods used to determine fine root production in forest ecosystems: a simulation approach. Can J For Res 17:909-912

Larsen MJ, Jurgensen MF, Harvey AE (1978) N_2 fixation associated with wood decayed by some common fungi in western Montana. Can J For Res 8:341-345

Larsen MJ, Jurgensen MF, Harvey AE (1982) N_2 fixation in brown-rotted soil wood in an intermountain cedar-hemlock ecosystem For Sci 28:292-296

Laskowski R, Berg B (1993) Dynamics of mineral nutrients and heavy metals in decomposing forest litter. Scand J For Res 8:446-456

Laskowski R, Berg B, Johansson M, McClaugherty C (1995) Release pattern for potassium from decomposing forest leaf litter. Long-term decomposition in a Scots pine forest XI. Can J Bot 73:2019-2027

Leatham GF, Kirk TK (1983) Regulation of lignolytic activity by nutrient nitrogen in white-rot basidiomycetes. FEMS Microbiol Lett 16:65-67

Lee PC, Crites S, Neitfeld M, Van Nguyen H, Stelfox JB (1997) Characteristics and origins of deadwood material in aspen-dominated boreal forests. Ecol Appl 7:691-701

Lewis DH (1980) Boron, lignification and the origin of vascular plants - a unified hypothesis. New Phytol 84:209-229

Lindbeck MR, Young JL (1965) Polarography of intermediates in the fixation of nitrogen by p-quinone-aqueous ammonia systems. Anal Chim Acta 32:73-80

Lindeberg G (1944) Über die Physiologie ligninabbauender Bodenhymenomyzeten. Symbol Botan Upsal VIII/2, 183 pp

Linkins AE, Sinsabaugh RL, McClaugherty CA, Melillo JM (1990) Cellulase activity on decomposing leaf litter in microcosms. Plant Soil 123:17-25

Liski J, Ilvesniemi H, Mäkelä A, Westman CO (1999) CO_2 emissions from soil in response to climatic warming are overestimated - the decomposition of old soil organic matter is tolerant to temperature. Ambio 28:171-174

Liu S-Y, Freyer AJ, Minard RD, Bollag J-M (1985) Enzyme-catalyzed complex-formation of amino esters and phenolic humus constituents. Soil Sci Soc Am J 49:337-342

Lockaby BG, Miller JH, Clawson RG (1995) Influences of community composition on biogeochemistry of loblolly pine (*Pinus taeda*) systems. Am Midl Nat 134:176-184

Lõhmus K, Ivask M (1995) Decomposition and nitrogen dynamics of fine roots of Norway spruce (*Picea abies* (L.) Karst.) at different sites. Plant Soil 168/169:89-94

Lousier JD, Parkinson D (1976) Litter decomposition in a cool temperate deciduous forest. Can J Bot 54:419-436

Lueken H, Hutcheon WL, Paul EA (1962) The influence of nitrogen on the decomposition of crop residues in the soil. Can J Soil Sci 42:276-288

Lundmark J-E Johansson MB (1986) Markmiljön i gran- och björkbestånd. Sver Skogs-vårdsförbunds Tidskr 2:31-37 (in Swedish)

MacMillan PC (1988) Decomposition of coarse woody debris in an old-growth Indiana forest. Can J For Res 18:1353-1362

Magill AH, Aber JD (1998) Long-term effects of experimental nitrogen additions on foliar litter decay and humus formation in forest ecosystems. Plant Soil 203:301-311

Mälkönen E (1974) Annual primary production and nutrient cycling in some Scots pine stands. Commun Inst For Fenn 84(5), 85 pp

Marra JL, Edmonds RL (1994) Coarse wood debris and forest floor respiration in an old-growth coniferous forest on the Olympic Peninsula, Washington, USA. Can J For Res 24:1811-1817

Marra JL, Edmonds RL (1996) Coarse wood debris and soil respiration in a clearcut on the Olympic Peninsula, Washington, USA. Can J For Res 26:1337-1345

Maser C, Anderson R, Cromack K, Williams JT, Martin RE (1979) Dead and down woody material. In: Thomas JW (ed) Wildlife habitats in managed forests, the Blue Mountains of Oregon and Washington. USDA For Serv Agric Handb 553, chap 6, pp 79-85

McCarthy BC, Bailey RR (1997) Distribution and abundance of coarse woody debris in managed forest landscape of the central Appalachians. Can J For Res 24:1317-1329

McClaugherty CA (1983) Soluble polyphenols and carbohydrates in throughfall and leaf litter decomposition. Acta Oecol 4:375-385

McClaugherty C Berg B (1987) Cellulose, lignin and nitrogen levels as rate regulating factors in late stages of forest litter decomposition. Pedobiologia 30:101-112

McClaugherty CA, Aber JD, Melillo JM (1982) The role of fine roots in the organic matter and nitrogen budgets of two forested ecosystems. Ecology 63:1481-1490

McClaugherty CA, Aber JD, Melillo JM (1984) Decomposition dynamics of fine roots in forested ecosystems. Oikos 42:378-386

McClaugherty CA, Pastor J, Aber JD, Melillo JM (1985) Forest litter decomposition in relation to soil nitrogen dynamics and litter quality. Ecology 66:266-275

McClellan MH, Bormann BT, Cromack K (1990) Cellulose decomposition in southeast Alaskan forests: effects of pit and mound microrelief and burial depth. Can J For Res 20:1242-1246

McHale PJ, Mitchell MJ, Bowles FP (1998) Soil warming in a northern hardwood forest: trace gas fluxes and leaf litter decomposition. Can J For Res 28:1365-1372

Means JE, MacMillan PC, Cromack K (1992) Biomass and nutrient content of Douglas-fir logs and other detrital pools in an old-growth forest, Oregon, USA. Can J For Res 22:1536-1546

Meentemeyer V (1978) Macroclimate and lignin control of litter decomposition rates. Ecology 59:465-472

Meentemeyer V (1984) The geography of organic decomposition rates. Ann Assoc Am Geogr 74:551-560

Meentemeyer V, Berg B (1986) Regional variation in mass-loss of *Pinus sylvestris* needle litter in Swedish pine forests as influenced by climate and litter quality. Scand J For Res 1:167-180

Meentemeyer V, Box E, Thompson R (1982) World patterns of terrestrial plant litter production. BioScience 32:125-128

Melillo JM, Aber JD, Muratore JF (1982) Nitrogen and lignin control of hardwood leaf litter dynamics in forest ecosystems. Ecology 63:621-626

Melillo JM, Aber JD, Linkins AE, Ricca A, Fry B, Nadelhoffer KJ (1989) Carbon and nitrogen dynamics along the decay continuum: plant litter to soil organic matter. In: Clarholm M, Bergström L (eds) Ecology of arable lands. Kluwer, Dordrecht, pp 53-62

Miller H (1979) The nutrient budgets of even-aged forests. In: Ford ED, Malcolm DC, Atterson J (eds) The ecology of even-aged forest plantations. Proc Meeting Division I. Internat Union For Res Org, Edinburgh, Sept 1978, pp 221-238

Miller HG, Miller JD (1976) Analysis of needle fall as a means of assessing nitrogen status in pine. Forestry 49:57-61

Mindermann G (1968) Addition, decomposition, and accumulation of organic matter in forests. J Ecol 56:355-362

Møller J, Miller M, Kjøller A (1999) Fungal-bacterial interaction on beech leaves: influence on decomposition and dissolved organic carbon quality. Soil Biol Biochem 31:367-374

Monleon VJ, Cromack K (1996) Long-term effects of prescribed burning on litter decomposition and nutrient release in ponderosa pine stands in central Oregon. For Ecol Manage 81:143-152

Moorehead DL, Sinsabaugh RL, Linkins AE, Reynolds JF (1996) Decomposition processes: modeling approaches and applications. Sci Total Environ 183:137-149

Muller RN, Liu Y (1991) Coarse woody debris in an old-growth forest on the Cumberland Plateau, southeastern Kentucky. Can J For Res 21:1567-1572

Musha Y, Goring DAI (1975) Distribution of syringyl and guaiacyl moieties in hardwoods as indicated by ultraviolet microscopy. Wood Sci Technol 9:45-58

Nadelhoffer KJ, Aber JD, Melillo JM (1985) Fine roots, net primary production, and soil nitrogen availability: a new hypothesis. Ecology 73:1377-1390

Næsset E (1999) Decomposition rate constants of *Picea abies* logs in southeastern Norway. Can J For Res 29:372-381

Newsham KK, Boddy L, Frankland JC, Ineson P (1992a) Effects of dry deposited sulphur dioxide on fungal decomposition of angiosperm tree leaf litter III. Decomposition rates and fungal respiration. New Phytol 122:127-140

Newsham KK, Frankland JC, Boddy L, Ineson P (1992b) Effects of dry deposited sulphur dioxide on fungal decomposition of angiosperm tree leaf litter I. Changes in communities of fungal saprotrophs. New Phytol 122:97-11040

Newsham KK, Ineson P, Boddy L, Frankland JC (1992c) Effects of dry deposited sulphur dioxide on fungal decomposition of angiosperm tree leaf litter II. Chemical content of leaf litters. New Phytol 122:111-125

Nikolov N, Helmisaari H (1992) Silvics of the circumpolar boreal forest tree species. In: Shugart HH, Leemans R, Bonan GB (eds) A systems analysis of the global boreal forest. Cambridge Univ Press, Cambridge, pp 13-84

Nilsson T, Daniel G, Kirk TK, Obst JR (1989) Chemistry and microscopy of wood decay by some higher ascomycetes. Holzforschung 43:11-18

Nömmik H, Möller G (1981) Nitrogen recovery in soil and needle biomass after fertilization of a Scots pine stand, and growth responses obtained. Stud Forest Suec 159, 37 pp

Nömmik H, Vahtras K (1982) Retention and fixation of ammonium and ammonia in soils. In: Stevenson FJ (ed) Nitrogen in agricultural soils. Agronomy monographs, no 22. Agron Soc Am, Madison, WI, pp 123-171

Norden B, Berg B (1990) A non-destructive method (solid state 13C-NMR) determining organic-chemical components in decomposing litter. Soil Biol Biochem 22:271-275

Olson JS (1963) Energy storage and the balance of producers and decomposers in ecological systems. Ecology 44:322-331

Olsson B, Bengtsson J, Lundkvist H, Staaf H, Rosen K (1996) Carbon and nitrogen in coniferous forest soils after clear-felling and harvests of different intensity. For Ecol Manage 82:19-32

O'Neill RV, Harris WF, Ausmus BS, Reichle DE (1975) A theoretical basis for ecosystem analysis with particular reference to element cycling. In: Howell FG, Gentry JB, Smith

MH (eds) Proceedings of the symposium on mineral cycling in southeastern ecosystems. US Dept Comm, Springfield, pp 28-40

Onega TL, Eickmeier WG (1991) Woody detritus inputs and decomposition kinetics in a southern temperate deciduous forest. Bull Torr Bot Club 118:52-57

Ono Y (1998) A study on the initial decomposition process of needle litter in a *Chamaecyparis obtusa* forest. Masters Thesis, Kyoto Univ, Kyoto, Japan (in Japanese)

Osono T, Takeda H (2001) Organic chemical and nutrient dynamics in decomposing beech leaf litter during 3-year decomposition process in a cool temperate deciduous forest in Japan. Ecol Res 16:649-670

Ovington JD (1959) The circulation of minerals in plantations of *Pinus sylvestris* L. Ann Bot NS 23:229-239

Panikov NS (1999) Understanding and prediction of soil microbial community dynamics under global change. Appl Soil Ecol 11:161-176

Panshin AJ, de Zeeuw C (1980) Textbook of wood technology, 4th edn. McGraw-Hill, New York, 722 pp

Paramesvaran N, Liese W (1982) Ultrastructural localization of wall components in wood cells. Holz Roh Werkst 40:145-155

Pastor J, Aber JD, McClaugherty CA, Melillo JM (1982) Geology, soils, and vegetation of Blackhawk Island, Wisconsin. Am Midl Nat 108:265-277

Pastor J, Aber JD, McClaugherty CA, Melillo JM (1984) Aboveground production and N and P cycling along a nitrogen mineralization gradient on Blackhawk Island, Wisconsin. Ecology 65:256-268

Paul E (1984) Dynamics of organic matter in soils. Plant Soil 76:75-285

Perala DA, Alban DH (1982) Rates of forest floor decomposition and nutrient turnover in aspen, pine, and spruce stands on two soils. US Dept Agric, For Serv, N Cent For Exp Stn, Res Pap NC-227, 5 pp

Perez J, Jeffries TW (1992) Roles of manganese and organic acid chelators in regulating lignin degradation and biosynthesis of peroxidases by *Phanerochaete chrysosporium*. Appl Environ Microbiol 58:2402-2409

Persson H (1980) Spatial distribution of fine-root growth, mortality and decomposition in a young Scots pine stand in Central Sweden. Oikos 34:77-87

Persson T, Karlsson PS, Seyferth U, Sjöberg RM, Rudebeck A (2000) Carbon mineralization in European forest soils. In: Schulze E-D (ed) Carbon and nitrogen cycling in European forest ecosystems. Ecological studies, vol 142. Springer, Berlin Heidelberg New York, pp 257-275

Persson T, Bååth E, Clarholm M, Lundkvist H, Söderström B, Sohlenius B, (1980) Trophic structure, biomass dynamics and carbon metabolism of soil organisms in a Scots pine forest. Ecol Bull (Stockholm) 32:419-462

Preston CM, Sollins P, Sayer BG (1990) Changes in organic components for fallen logs in old-growth Douglas-fir forests monitored by ^{13}C nuclear magnetic resonance spectroscopy. Can J For Res 20:1382-1391

Poinsot-Balaguer N, Tabone E (1995) Impact of chronic gamma irradiation on the litter decay of a mixed Mediterranean forest in Cadarache - France - microarthropod's response. Pedobiologia 39:344-350

Prescott CE (1995) Does nitrogen availability control rates of litter decomposition in forests? Plant Soil 168/169:83-88

Prescott CE, Blevins LL, Staley CL (2000) Effects of clear-cutting on decomposition rates of litter and forest floor in forests of British Columbia. Can J For Res 30:1-7

Pyle C, Brown MM (1998) A rapid system of decay classification for hardwood logs of the eastern deciduous forest. J Torr Bot Soc 125:237-245

Pyle C, Brown MM (1999) Heterogeneity of wood decay classes within hardwood logs. For Ecol Manage 114:253-259

Raich JW, Nadelhoffer KJ (1989) Belowground carbon allocation in forest ecosystems; global trends. Ecology 70:1346-1354

Rapp M, Leonardi S (1988) Evolution de la litiere au sol au cours d'une annee dans un taillis de chene vert (*Quercus ilex*). Pedobiologia 32:177-185 (in French)

Rashid GH, Schaefer R (1988) Seasonal variation in the nitrogen mineralization and mineral nitrogen accumulation in two temperate forest soils. Pedobiologia 31:381-390

Rehfüss KE (1990) Waldböden, Entwicklung, Eigenschaften und Nutzung, 2nd edn. Pareys Studientexte 29. Parey, Hamburg, 294 pp (in German)

Reid ID, Seifert KA (1982) Effect of an atmosphere of oxygen on growth, respiration, and lignin degradation by white-rot fungi. Can J Bot 60:252-260

Reurslag AM, Berg B (1993) Rapport över litteraturstudie rörande mängd och kemisk sammansättning av fallförna samt mängd av organiskt material i skogsmark. Vattenfalls rapportserie No UB 1993/2, 110 pp (in Swedish, English summary)

Robertson PA, Bowser YH (1999) Coarse woody debris in mature *Pinus ponderosa* stands in Colorado. J Torr Bot Soc 126(3):255-267

Roo-Zielinska E, Solon J (1997) Effect of geographical location on species composition, vegetation structure, diversity and phytoindicative characteristics in pine forests. Environ Poll 98:347-360

Roo-Zielinska E, Solon J (1998) A geobotanical characteristic and analysis of the range of forest communities at study sites along climatic in pine and mixed pine forests along a climatic (52o N, 12-32oE) and a "Silesian" transect. In: Breymeyer A, Roo-Zielinska E (eds) Pine forests in central European gradient of continentality and pollution - geoecological studies. Docum Geogr 13:79-98

Rose RE, Lisse MW (1916) The chemistry of wood decay paper I - introductory. J Ind Eng Chem 9:284-287

Rustad LE, Fernandez IJ (1998) Soil Warming: consequences for foliar litter decay in a spruce-fir forest in Maine, USA. Soil Sci Soc Am J 62:1072-1080

Rutigliano FA, Virzo De Santo A, Berg B, Alfani A, Fioretto A (1996) Lignin decomposition in decaying leaves of *Fagus sylvatica* L. and needles of *Abies alba* Mill. Soil Biol Biochem 28:101-106

Ryan MG, Melillo JM, Ricca A (1990) A comparison of methods for determining proximate carbon fractions of forest litter. Can J For Res 20:166-171

Saka S, Thomas RJ (1982a) Evaluation of the quantitative assay of lignin distribution by SEM-EDXA technique. Wood Sci Technol 16:1-18

Saka S, Thomas RJ (1982b) A study of lignification in loblolly pine tracheids by the SEM-EDXA technique. Wood Sci Technol 16:167-179

Salonius PO (1983) Effects of organic-mineral soil mixtures and increasing temperature on the respiration of coniferous raw humus material. Can J For Res 13:102-107

Santantonio D, Hermann RK (1985) Standing crop, production, and turnover of fine roots on dry, moderate, and wet sites of mature Douglas-fir in western Oregon. Ann Sci For 42:113-142

Schiffman PM, Johnson WC (1989) Phytomass and detrital carbon storage during forest regrowth in the southeastern United States Piedmont. Can J For Res 19:67-78

Schlesinger WH (1977) Carbon balance in terrestrial detritus. Annu Rev Ecol Syst 8:51-81

Schlesinger WH, Andrews JA (2000) Soil respiration and the global carbon cycle. Biogeochemistry 48:7-20

Schlesinger WH, Hasey MM (1981) Decomposition of chaparral shrub foliage; losses of organic and inorganic nutrients from deciduous and evergreen leaves. Ecology 62:762-774

Schowalter TD (1992) Heterogeneity of decomposition and nutrient dynamics of oak (*Quercus*) logs during the first 2 years of decomposition. Can J For Res 22:161-166

Schowalter TD, Zhang YL, Sabin TE (1998) Decomposition and nutrient dynamics of oak (*Quercus* spp.) logs after five years of decomposition. Ecography 21:3-10

Schulze E-D, de Vries W, Hauhs M, Rosen K, Rasmussen L, Tamm CO, Nilsson J (1989) Critical loads for nitrogen deposition on forest ecosystems. Water Air Soil Pollut 48:451-456

Schmidt EL, Ruschmeyer OR (1958) Cellulose decomposition in soil burial beds. I. Soil properties in relation to cellulose degradation. Appl Microbiol 6:108-114

Scott NA, Cole CV, Elliott ET, Huffman SA (1996) Soil textural control on decomposition and soil organic matter dynamics. Soil Sci Soc Am J 60:1102-1109

Silver WL, Miya RK (2001) Global patterns in root decomposition: comparisons of climate and litter quality effects. Oecologia 129:407-419

Singh JS, Gupta SR (1977) Plant decomposition and soil respiration in terrestrial ecosystems. Bot Rev 43:449-528

Singh AP, Nilsson T, Daniel GF (1987) Ultrastructure of the attack of the wood of two high lignin tropical hardwood species, *Alstonia scholaris* and *Homalium foetidum*, by tunneling bacteria. J Inst Wood Sci 11:237-249

Staaf H (1982) Plant nutrient changes in beech leaves during senescence as influenced by site characteristics. Acta Oecol/Oecol Plant 3:161-170

Staaf H, Berg B (1977) A structural and chemical description of litter and humus in the mature Scots pine stand at Ivantjärnsheden. Swed Conif For Proj Int Rep 65, 31 pp

Staaf H, Berg B (1981) Plant litter input. In: Clark FE, Rosswall T (eds) Terrestrial nitrogen cycles. Processes, ecosystem strategies and management impacts. Ecol Bull (Stockh) 33:147-162

Staaf H, Berg B (1982) Accumulation and release of plant nutrients in decomposing Scots pine needle litter. Long-term decomposition in a Scots pine forest II. Can J Bot 60:1561-1568

Staaf H, Berg B (1989) Förnanedbrytning som variabel för miljökvalitetskontroll i terrester miljö. I. Substraturval och inledande studier. Statens Naturvårdsverk Rapport Nr 3591, 62 pp (in Swedish)

Stevenson FJ (1982) Humus chemistry. Genesis, composition, reactions. Wiley, New York, 443 pp

Stewart GH, Burrows LE (1994) Coarse woody debris in old-growth temperate beech (*Nothofagus*) forests of New Zealand. Can J For Res 24:1989-1996

Stone JN, MacKinnon A, Parminter JV, Lertzman (1998) Coarse woody debris decomposition documented over 65 years on southern Vancouver Island. Can J For Res 28:788-793

Strömgren M (2001) Soil-surface CO_2 flux and growth in a boreal Norway spruce stand. Effects of soil warming and nutrition. Doctoral dissertation, Acta Universitatis Agriculturae Sueciae, Silvestria 220, 44 pp

Strojan CL (1978) Forest leaf litter decomposition in the vicinity of a zinc smelter. Oecologia 32:203-212

Sturtevant BR, Bissonette JA, Long JN, Roberts DW (1997) Coarse woody debris as a function of age, stand structure, and disturbance in boreal Newfoundland. Ecol Appl 7:702-712

Swedish Natl Survey of Forest Soils and Vegetation (1983-1987) Dept of Forest Soils, Swed Univ Agric Sci, Uppsala

Swift MJ (1977) The ecology of wood decomposition. Sci Prog Oxf 64:175-199

Swift MJ, Heal OW, Anderson JM (1979) Decomposition in terrestrial ecosystems. Blackwell, Oxford

Tamm CO (1991) Nitrogen in terrestrial ecosystems. Questions of productivity, vegetational changes, and ecosystem stability. Ecological studies, vol 81. Springer, Berlin Heidelberg New York, 115 pp

Tamm CO (1999) Optimum nutrition and nitrogen saturation in Scots pine stands. Stud Forest Suec 206:1-126

Tamm CO, Nilsson Å, Wiklander G (1974) The optimum nutrition experiment Lisselbo: a brief description of an experiment in a young stand of Scots pine (*Pinus sylvestris* L.). Rapporter och uppsatser no 18. Institutionen för växtekologi och marklära, Skogshögskolan, Stockholm, 25 pp

Taylor BR, Parkinson D, Parsons WFJ (1989) Nitrogen and lignin content as predictors of litter decay rates: a microcosm test. Ecology 70:91-104

Tenney FG, Waksman SA (1929) Composition of natural organic materials and their decomposition in the soil. IV. The nature and rapidity of decomposition of the various organic complexes in different plant materials, under aerobic conditions. Soil Sci 28:55-84

Thornthwaite CW, Mather JR (1957) Instructions and tables, for computing potential evapotranspiration and the water balance. Publ Climatol 10:185-311

Tien M, Kirk TK (1984) Lignin-degrading enzyme from *Phanerochaete chrysosporium*: purification, characterization, and catalytic properties of a unique H_2O_2- requiring oxygenase. Proc Natl Acad Sci USA 81:2280-2284

Torsvik VL, Goksøyr J, Daae FL (1990) High diversity of DNA in soil bacteria. Appl Environ Microbiol 56:782-787

Townsend AR, Vitousek M, Trumbore S (1995) Soil organic matter dynamics along gradients in temperature and land use on the island of Hawaii. Ecology 76:721-733

Townsend AR, Vitousek P, Desmarais DJ, Tharpe A (1997) Soil carbon pool structure and temperature sensitivity inferred using CO_2 and $^{13}CO_2$ incubation fluxes from five Hawaiian soils. Biogeochemistry 38:1-17

Tsuneda A, Thorn RG (1995) Interactions of wood decay fungi with other microorganisms, with emphasis on the degradation of cell walls. Can J Bot 73:S1325-S1333

Turner J, Long JN (1975) Accumulation of organic matter in a series of Douglas-fir stands. Can J For Res 5:681-690

Tyrell LE, Crow TR (1994) Dynamics of deadwood in old-growth hemlock-hardwood forests of northern Wisconsin and Northern Michigan. Can J For Res 24:1672-1683

Ulrich B (1981) Teoretische Betrachtung des Ionenkreislaufs in Waldökosystemen. Z Pflanzenern Bodenkd 144:289-305 (in German)

UNECE/FAO (2000) Forest Resources of Europe, CIS, North America, Australia, Japan, and New Zealand, United Nations Economic Commission for Europe and Food and Agricultural Organization. Geneva Timber and Forest Studies Papers, no 17. United Nations, Geneva, 467 pp

Unestam T (1991) Water repellency, mat formation, and leaf-stimulated growth of some ectomycorrhizal fungi. Mycorrhiza 1:13-20

Upadhyay VP, Singh JS (1985) Nitrogen dynamics of decomposing hardwood leaf litter in a central Himalayan forest. Soil Biol Biochem 17:827-830

Van Cleve K (1974) Organic matter quality in relation to decomposition. In: Holding AJ, Heal OW, MacLean SF Flanagan PW (eds) Soil organisms and decomposition in Tundra. Tundra Biome Steering Committee, Stockholm, pp 311-324

Van Cleve K, Oechel WC, Hom JL (1990) Response of black spruce (*Picea mariana*) ecosystems to soil temperature modification in interior Alaska, USA. Can J For Res 20:1530-1535

Van Soest PJ (1963) Use of detergents in the analysis of fibrous feeds. II. A rapid method for the determination of fiber and lignin. J Assoc Off Agric Chem 46(5):829-835

Verburg PSJ, Van Loon WKP, Lükewille (1999) The CLIMEX soil-heating experiment: soil response after 2 years of treatment. Biol Fertil Soils 28:271-276

Viljoen JA, Fred ED, Peterson WH (1926) The fermentation of cellulose by thermophilic bacteria. J Agric Sci 16:1-17

Vogt KA, Vogt DJ, Gower ST, Grier CC (1990) Carbon and nitrogen interactions for forest ecosystems. In: Persson H (ed) Above- and below-ground nitrogen interactions in forest trees in acidified soils. Air pollution report 32. Commission of the European Communities, Directorate-General for Science, Research and Development. Environment Research Programme, Brussels, Belgium, pp 203-235

Vogt KA, Vogt DJ, Bloomfield J (1991) Input of organic matter to the soil by tree roots. In: Persson H, McMichael BL (eds) Plant roots and their environments. Elsevier, Amsterdam, pp 171-190

Vogt KA, Vogt DJ, Palmiotto PA, Boon P, O'Hara J, Asbjornsen H (1996) Review of root dynamics in forest ecosystems grouped by climate, climatic forest type and species. Plant Soil 187:159-219

von Liebig J (1847) Chemistry in its applications to agriculture and physiology. Taylor and Walton, London, 418 pp

Waksman SA (1936) Humus. Williams and Wilkins, Baltimore

Waksman SA, Reuszer HW (1932) On the origin of the uronic acids in the humus of soil. peat, and composts. Soil Sci 33:135-151

Waksman SA, Tenney FG, Stevens KR (1928) The role of microorganisms in the transformation of organic matter in forest soils. Ecology 9:126-144

Wang Y, Amundson R, Trumbore S (1996) Radiocarbon dating of soil organic matter. Quat Res 45:282-288

Wardle DA, Zachrisson O, Hörnberg G, Gallet C (1997) The influence of island area on ecosystem properties. Science 277:1296-1299

Watson RT, Meira Filho LG, Sanhueza E, Janetos A (1992) Greenhouse gases: sources and sinks. In: Houghton JT, Callander BA, Varney SK (eds) Climate change 1992. The supplementary report to the IPCC scientific assessment. Cambridge Univ Press, Cambridge, pp 25-46

Wieder RK, Lang GE (1982) A critique of the analytical methods used in examining decomposition data obtained from litter bags. Ecology 63:1636-1642

Wiegel J, Dykstra M (1984) *Clostridium thermocellum*: adhesion and sporulation while adhered to cellulose and hemicellulose. Appl Microbiol Biot 20:59-65

Wolter KE, Highley TL, Evans FJ (1980) A unique polysaccharide- and glycoside-degrading enzyme complex from the wood decay fungus *Poria placenta*. Biochem Biophys Res Commun 97:1499-1504

Woodwell GM, Marples TG (1968) The influence of chronic gamma irradiation on production and decay of litter and humus in an oak-pine forest. Ecology 49:456-465

Woodwell GM, Mackenzie FT, Houghton RA, Apps M, Gorham E, Davidson E (1998) Biotic feedbacks in the warming of the earth. Clim Change 40:495-518

Wookey PA, Ineson P (1991) Chemical changes in decomposing forest litter in response to atmospheric sulphur dioxide. J Soil Sci 42:615-628

Wookey PA, Ineson P, Mansfield TA (1991) Effects of atmospheric sulphur dioxide on microbial activity in decomposing forest litter. Agric Ecosyst Environ 33:263-28

Worrall JJ and Wang CJK (1991) Importance and mobilization of nutrients in soft rot of wood. Can J Microbiol 37:864-868

Worrall JJ, Anagnost SE, Wang CJK (1991) Conditions for soft-rot of wood. Can J Microbiol 37:869-874

Worrall JJ, Anagnost SE, Zabel RA (1997) Comparison of wood decay among diverse lignicolous groups. Mycologia 89:199-219

Wright RF, Tietema A (1995) Ecosystem response to 9 years of nitrogen addition at Sogn-dal, Norway. For Ecol Manage 71:133-142

Wright RF, Roelofs JGM, Bredemeier M, Blanck K, Boxman AW, Emmett BA, Gundersen P, Hultberg H, Kjønaas OJ, Moldan F, Tietema A, van Breeman N, van Dijk HFG (1995) NITREX: responses of coniferous forest ecosystems to experimentally changed deposition of litter. For Ecol Manage 71:163-169

Yin X (1999) The decay of forest woody debris: numerical modeling and implications based on some 300 data cases from North America. Oecologia 121:81-98

Yin XW, Perry JA, Dixon RK (1989) Influence of canopy removal on oak forest floor de-composition. Can J For Res 19:204-214

Zachrisson O (1977) Influence of forest fires on the northern Swedish boreal forest. Oikos 29:22-32

Zanetti S, Hartwig UA (1997) Symbiotic N2 fixation increases under elevated atmospheric pCO$_2$ in the field. Acta Oecol 18:285-290

Zhang Q, Liang Y (1995) Effects of gap size on nutrient release from plant litter decompo-sition in a natural forest ecosystem. Can J For Res 25:1626-163

Index

Druck: Strauss Offsetdruck, Mörlenbach
Verarbeitung: Schäffer, Grünstadt